HIGH-TECH FANTASIES

High-tech Fantasies goes beyond the normal 'policy evaluation' to examine the underlying assumptions which science parks embody about science and society. It is argued that science parks are founded on a notion of scientific production and industrial innovation which is not only technically inappropriate but also intrinsically socially divisive. Moreover, the spatial form and symbolic spatial content of science parks – close to academe, separated from physical production and with specific design characteristics – further increases their tendency to promote (and depend upon) social polarisation. In the UK, for example, this polarisation takes a precise geographical form, exemplified in the clustering of 'high tech' in the semi-rural regions of the south and east of England.

The book focuses on the mutual relationship between the forms of scientific production, social structures and geographical inequality. It dismantles the popular concept of science parks and presents an alternative conceptualisation from which the real implications of science-park developments can be drawn. This alternative is used to analyse both the room for manoeuvre available to those currently involved in developing science parks and the ultimate constraints imposed by their assumptions.

HIGH-TECH FANTASIES

Science parks in society, science and space

Doreen Massey, Paul Quintas and David Wield

London and New York

First published 1992
by Routledge
11 New Fetter Lane, London EC4P 4EE

Simultaneously published in the USA and Canada
by Routledge
a division of Routledge, Chapman and Hall, Inc.
29 West 35th Street, New York, NY 10001

Typeset by GilCoM Ltd, Mitcham, Surrey.
Printed and bound in Great Britain by
Biddles Ltd, Guildford and King's Lynn

British Library Cataloguing in Publication Data
Massey, Doreen 1944–
High-tech fantasies: science parks in society, science and space.
1. Great Britain. Science parks
I. Title II. Quintas, Paul III. Wield, David
306.45
ISBN 0–415–01338–0
ISBN0–415–01339–9 pbk

Library of Congress Cataloging-in-Publication Data
Massey, Doreen B.
High-tech fantasies: science parks in society, science, and space
Doreen Massey, Paul Quintas, and David Wield.
p. cm.
Includes bibliographical references and index.
ISBN 0–415–01338–0 (HB). – ISBN 0–415–01339–9 (PB)
1. Research parks – Great Britain – Social aspects. 2. Research parks – Great Britain – Economic aspects. I. Quintas, Paul, 1947–
II. Wield, David. III. Title.
T175.7.M37 1992
303.48'3–dc20 91 – 3191
CIP

CONTENTS

List of figures vii
List of tables viii
Preface xi
Acknowledgements xiv

1 THE ISSUES 1

2 SCIENCE PARKS IN THEIR OWN TERMS 13
The popular conceptualisation 13
Definition, and a thumbnail sketch 13
Postulated effects 20
Evaluating science parks on their own terms 30
The formation of new start-up firms 30
Facilitation of R&D links and technology transfer between host academic institutions and park firms 34
Employment 'creation' 40
The level of technology 43
Conclusions and further questions 51

3 SCIENTIFIC AND TECHNOLOGICAL DIVISIONS 56
The linear model of innovation 56
The model: foundation concept of science parks 56
The roots, and cultural specificity 60
Difficulties with the linear model 72
Science parks and the linear model 72
Critique of the model 76

4 ON THE PARK – A VISION OF THE FUTURE OF WORK? 86
The drive to Cambridge 87
Working on the park 91
Flexible working time 92

Organisational structure, status and functional flexibility 96
Flexible pay? 101
Trade unions 104
The flexibility debate 106
Problems with the future: on the park and beyond 108
Internal ambiguities 108
Exclusivity 109

5 SCIENCE PARKS AND SOCIAL STRUCTURE 115
The class location of scientists and technologists 116
Some history 116
Debate 122
Recent changes in the social location of scientists and technologists 130
Traditionally high-status jobs 130
A new hierarchy? 138
Reflections 144
Status and social power 148

6 SCIENCE PARKS AND LOCAL ECONOMIES 163
The economic history 165
Aston 165
Cambridge 170
Park links 178
The international scene 178
Links with the local economy 184
Labour markets 187
Research links 195
Other contrasts 196
Reflections 203

7 SCIENCE PARKS AND THE PUBLIC PURSE 206
Individualism and entrepreneurship 206
Public funds and private accumulation 209
Science parks and public funding 209
Public funds and production on the parks 226
Public–private relations and uneven development 235

8 CONCLUSIONS 239

References 251
Index 261

FIGURES

2.1	Science parks eligible to join UKSPA (end 1988)	15
2.2	Leading edge? Examples of more sophisticated establishments	49
2.3	Leading edge? Examples of less sophisticated establishments	50
3.1	The linear innovation model	57
3.2	Invention, innovation and diffusion as a linear process	77
3.3	Product and process innovation	77
3.4	The process of technological innovation – an interactive model	82
3.5	Chain-linked model of innovation	84
5.1	Wright's (1976) class framework	117
5.2	Wright's (1985) typology of class locations in capitalist society	125
6.1	The spatial structures of firms with activities on science parks: a comparison of Aston and Cambridge	180
7.1	Industrial rents: the south-east 'premium'	211
7.2	High-technology rental values	216
7.3	High-tech and other industrial property	227

TABLES

2.1	Science parks by year of opening	16
2.2	The North–South divide in science-park development	16
2.3	Science-park tenants, 1986–8	18
2.4	Sources of investment in UK science parks	19
2.5	Park tenants and employment, 1985–90	20
2.6	Employment in science parks	20
2.7	Summary of science-park objectives	21
2.8	How many park units are new start-ups?	31
2.9	Age of surveyed establishments on science parks	33
2.10	North–South differences in new start-ups	34
2.11	Links with the local academic institution	39
2.12	Previous location of relocated park establishments	42
2.13	Employment on science parks, by occupational category	43
2.14	Industrial sectors of science-park establishments	44
2.15	Main sectors of science-park establishments	44
2.16	Information technology concentration: Cambridge and Aston compared with park mean	45
2.17	Principal on-site activities of establishments	46
2.18	Patenting activity of park and off-park establishments	47
2.19	New products and patent activity compared	48

TABLES

4.1	Time flexibility on science parks	93
4.2	Functional flexibility	98
4.3	Participation in decision-making	99
4.4	The payment of overtime	102
4.5	Trade-union membership	104
4.6	Women working on Aston and Cambridge	110
4.7	Women and part-time work	112
5.1	Employer category of first-degree science graduates entering permanent employment in the UK	134
5.2	International comparison of salaries of departmental directors	139
5.3	Professional engineers', scientists' and technologists' employment, by selected sector and size of establishment, 1985	143
6.1	Average wage and salary levels in Cambridge	192
6.2	Locally recruited personnel, Cambridge	194
7.1	Public and private-sector funding for science parks and high-tech developments: geographical contrasts	210
7.2	Prime industrial rents: the contrast between London and the rest of the country	212
7.3	Concentration of infrastructural investment in parks	218
7.4	Sources of finance for firms	230
7.5	Sources of finance mentioned as being 'most important', Aston and Cambridge	232
7.6	Sources of finance for firms, Aston and Cambridge	234

PREFACE

This book began life at a seminar of the Open University Technology Policy Group in 1983. David Elliott, presenting his argument that 'science parks' might be more than a passing fad, but rather could be a significant 1980s change in science and technology policy (and, it emerged during the seminar, industrial policy), convinced two of us (Doreen Massey and David Wield) that social scientists and technologists should collaborate in investigating further. An Open University funded pilot project led to a grant from the Joint Committee of the ESRC–SERC and the appointment of a full-time researcher (Paul Quintas).

Since the early days of our research science parks in the UK have grown from a handful and a set of ideas and building plans. They have become very much a part of the vocabulary of a diverse range of property developers, management consultants, government ministers, local councillors and planners, and regional authorities, in the UK and internationally. As science parks developed so did the scope of our project. Over time, our focus on the study of science parks has allowed us to deepen our arguments on the relationship between science and industrial innovation, changing work patterns and changing divisions of labour and the regional geography of development. One major aim has been to show that collaboration between social and natural scientists can yield results of use to both and all the more useful to those interested in social and economic development.

The success of the research has depended on the support of literally hundreds of people. We are grateful to the ESRC–SERC Joint Committee, Open University Technology and Social Sciences Faculties, and Peat Marwick McLintock for funding for staff and research. David Elliott has been a collaborator for most of the project's history. The three authors have produced it together, everyone doing some of both data gathering and analysis and write-up. Paul Quintas took the major responsibility for data gathering, particularly from science parks, David Wield for broader science and technology issues, such as sci-tech careers, and Doreen Massey for local area information. The book has gone through various drafts, evenly divided in the early stages, with David Wield and particularly Doreen Massey producing the final draft. We stand together in taking responsibility for the final product. It is difficult to conceive of a tighter collaboration or of better support.

The United Kingdom Science Parks Association gave support in various ways. Charles Monck answered our early questions and took seriously our view that research collaboration at the level of data collection might lessen the plight of science-park managers and establishments as researchers descended on them in the mid-1980s. Much of our quantitative data collection came from a collaboration organised by UKSPA and involving the Centre for Urban and Regional Studies (David Storey and Pooran Wynarcyk) and Peat Marwick (Dick Porter and Rory Landman). At UKSPA Brian Worrall and Susan Cooke have helped in many ways. Pilot research was conducted by Sheila Cohen, Stuart Russell and Keith Dickson. Our in-depth research at Cambridge Science Park was supported by Henry Bennett and Lindy Beveridge. At Aston Science Park we received tremendous support from the staff of Birmingham Technology Ltd. Harry Nicholls and Viv Holland arranged our interviews and much more. Officers of the Economic Development Department of Birmingham City Council also gave valuable assistance. To those managements of twenty parks who gave us time and information we offer our thanks. A wide range of people in other local authorities and agencies gave help. Our particular thanks go to John Darwin, Bridget Pemberton and Gordon Dabinett. This long list does not exhaust those who gave of their time, particularly those who allowed themselves to be interviewed in establishments, and who gave information from their knowledge of science-park development. David Wield would like to acknowledge the support of the VTSS Program at Stanford University, particularly Jim Adams, Steve Kline, Robert McGinn and Walter Vincenti, and thank them for making his stay in 1989–90 so pleasurable, and also members of Geography and IURD at UC Berkeley.

PREFACE

This book was produced using relatively sophisticated technology with a significant division of labour. For instance, several computer systems from the Social Science and Technology Faculties had to be integrated. John Hunt produced various figures and maps. Our particular thanks go to Doreen Warwick, who co-ordinated the manuscript's production, and whose ability with a high-tech and constantly changing production system was invaluable. We would also like to acknowledge the help of Margaret Charters. Their high-tech accomplishments are certainly no fantasy.

ACKNOWLEDGEMENTS

The authors and publisher wish to thank the following who have kindly given permission for the use of copyright material: Robin Roy for Figure 3.4; Stephen Kline for Figure 3.5; Erik Olin Wright for Figures 5.1 and 5.2; Debenham Tewson and Chinnocks for Figures 7.1 and 7.3; John Henneberry for Table 7.1.

1

THE ISSUES

The empirical project which eventually led to this book seemed, when we began, quite straightforward. We were interested in science parks. In particular, we were interested in the relation between their scientific and technological content, their social structure and meaning and their spatial distribution and form.

What happened was precisely what academics warn their students against. The project blossomed and grew. It did so in part because of a concern with conceptualisation which led us in the end to rethink what is the real content of the term 'science park'. This took us in wider, and unexpected, directions. The project also grew because this apparently simple empirical object of study in fact increasingly linked in, as we explored it, to some major theoretical and political issues.

The chapters which follow, therefore, range very widely. They explore the place of scientists, technologists and engineers in the class structure of capitalist societies, and how that place is changing. They examine the social and scientific bases of the hype around high tech. They take up arguments about the nature of scientific research and industrial innovation. They examine the impact of some of the recent cuts in university research budgets. They look at the impact on, and of, geography in the new era of 'high tech', and the social significance of spatial symbolism. They examine the difficulties facing local and regional authorities in some parts of the country in devising local economic strategies to bring their areas into contact with a so-called sunrise future; and the impact of financial capital and the property development industry in reinforcing the North–South divide in the UK.

What holds these things together is a concern about the relationship (and the possible alternative relationships) between science, society and space. We argue that the current development of science parks, certainly in the UK, is symbolic of wider changes, and of a wider politics. Science parks bring together and interrelate particular ideologies and practices of scientific advance and industrial innovation; divisions of labour within society and their related social structures; and the geography of social and economic development.

It is, of course, not unusual to make links between these aspects of society. There are large literatures on, for example, the social impact of technological change, the geography of particular social structures, or the spatial implications of high technology. In much of these literatures, however, the links between the pairs are one-way. 'Technology' has an impact on social structure, or on spatial organisation. Social change 'produces' spatial change. One of the arguments of this book is that all the links between these pairs are two-way, if sometimes unequally so. Social structures and political choices lie behind particular models of industrial innovation and even scientific procedures. Geography can have an impact on the meaning and real content of social structures. Thus it has been a central theme of much research in the urban and regional field since the 1970s that particular forms of uneven development are the product of particular forms of social development, and that the problems posed by the former thus require intervention in the latter. This argument is clearly supported by our study of science parks. But the study also indicates how spatial organisation is itself instrumental in the construction of social status and differentiation. Similarly, 'technology' is all too often seen in the social sciences as a *deus ex machina* which it is the job of social scientists to take as given, in order to evaluate its effects, both social and spatial. Yet on the other side of the great divide within academe natural scientists and technologists have increasingly been emphasising the social construction of technology itself, and of the structures and processes of scientific research and industrial innovation (Harding 1986; MacKenzie and Wajcman 1985; Bijker *et al.* 1989). Thus, not only may any attempt to change geographical outcomes depend on changing social processes and structures, but those social structures themselves are intimately bound up with (in the aspects of society we are considering here) socially constructed models of science and technology. Such models of science and technology may also have to be questioned before any serious alteration can be made to geographical uneven development. This book is the product of a joint effort by social and natural scientists and one of its key themes is precisely the *mutual* determination of the socio-spatial on the one hand and the scientific and the technological on the other.

These are general arguments. Within this framework, science parks in their archetypal form emerge as highly particular expressions of each of these relationships. They are, we argue, based on a rigid and not necessarily very productive model of invention and innovation; they are expressions of a highly hierarchical social structure which itself is intimately related to that founding model of innovation; and they are as geographical entities constructed around notions of difference and status which both reinforce the social hierarchy and further rigidify the technological model. An examination of science parks is thus in fact an exploration of some of the key current relationships between science, society and space.

Much of this book is about differences and divisions of labour in society; about how to conceptualise them, about how to break them down, and about the potential to reformulate them in a more egalitarian and socially productive way. The production of the argument by natural scientists and a geographer was itself an experiment in crossing C. P. Snow's great divide (he also crops up in the discussion) between the arts and the sciences, so strong in the United Kingdom. A basic thread of analysis which runs throughout the whole book is the intimate connection (which does not mean deterministic relation) between technical divisions of labour in the structures of scientific production and industrial innovation, divisions of labour in society more widely and the cultures of class, and spatial divisions of labour as different functions within the overall economy of a society take on particular geographical forms. There are generalisations one can make about each of these divisions of labour, and the relations between them, which probably apply to most industrial capitalist societies. But each society also exhibits its specificities, and here we investigate in particular the peculiarity of the British case. There is, we argue, a systematic connection between the culture of the nineteenth-century British (perhaps better English) upper middle class with its disdain for the ugly facts of manufacturing production, the dismal performance in scientific and technological terms of much of the British industrial economy, and the particular, and particularly sharp, form of 'North–South' divide exhibited by the new high-tech sectors of today.

One of our central concerns is with the *conceptualisation* of difference and in particular of divisions of labour. There is an important distinction to be made, both conceptually and in terms of political implications, between difference as simply descriptive but not mutually determining variation, and a set of differences which are mutually constitutive, where their definition is relational. Moreover, within the latter the differences may be egalitarian or inegalitarian. Throughout our investigation, whether we are dealing with divisions of labour within the practice of scientific investigation, the distribution of occupations in the economy into different levels of real skill content, or the spatial divisions of labour within the economy at large, the divides and differences which we encounter are mutually constitutive. The fact that there is a specified activity called 'applied science' implies the existence of one called 'basic' or 'pure' science. The fact that some people are experts in something implies that others are not. The fact that the south and east of England are the cradle of the United Kingdom's high-technology industry necessitates that other areas cannot be so. Moreover, these differences are not only mutually constitutive but in the particular cases we are examining they are unequally and hierarchically so. There is mental and manual labour, skilled and unskilled work, there are regions of research and development and parts of the country with none of these things.

There is also a further point: that some aspects of the present era of high technology are exacerbating many of these divides. Science parks are often held up as images of the future. Yet, as we show, their very construction – scientifically, socially and spatially – is based on notions of unequal, mutually constitutive difference. This is true in terms of the model of science on which they rest, the social content by which they define themselves, and the spatial symbolism through which they are physically designed. The construction of status is necessarily counterposed, in all these areas, to an absence of status, and is by definition not generalisable. Much of the hype surrounding 'high-tech', in other words, whether it relates to the division of labour, social status, or geographical location, is exclusive and at the expense of those excluded.

Science parks have been big news for some time now. In the United States of America, where they originated, and in the United Kingdom, where there are around forty, universities, local and regional public-sector bodies, and commercial agencies of many kinds have all been involved in their establishment. They are found in countries throughout Europe. In Madrid, in Catalonia, in the Basque country, the major regions of Spain each have one. France has its technopoles. There has even been talk of a string of them being developed down the Yugoslav coast of the Adriatic. Japan has developed its own variant of the phenomenon. In regions across the world, including many developing countries, local agencies of a variety of social and political hues pin their hopes on them (or, more usually, something like them) for capturing a place in a high-tech future.

It is important, therefore, simply because of the burgeoning nature of this phenomenon to analyse what it is all about. However, we argue, doing this adequately means analysis at two quite distinct levels. At one level it means simply taking a sober look at the empirical reality of science parks as they have developed so far. What kind of employment do they contain? How are they related to the local economies in which they are set, and to their growth and change? In what ways are science parks related to the holy grail of 'high technology'? To what extent have they fulfilled their own stated goals? This, in other words, is what most evaluations of policy do – though our conclusions are a bit different from some. However, second, we argue that it is also necessary to step back and examine science parks in their wider social context. Our argument at this second level is that in their archetypal form science parks condense a whole range of characteristics of current social developments. Although they are widespread, their quantitative importance is, as yet, limited. There were, by the beginning of the 1990s, only something over 14,000 people employed on them in the United Kingdom, for instance. But the significance of science parks goes beyond this quantitative evaluation. The bulk of the book, therefore, is concerned with examining these more fundamental issues.

The original foundations for the idea of science parks derived from the United States, from Stanford and Silicon Valley and from Boston–Cambridge and Route 128. In both these areas there was a combination of a major academic community and the rapid growth of high-technology industry. The empirical juxtaposition was assumed to be causal (and later, also, to be reproducible), and a story soon evolved and spread, built around some key components. There were tales of entrepreneurial professors, the spawning of new companies as bright ideas dreamt up in academe found favour in the market place, and were turned into commercial successes. There were venture capitalists prepared to take risks. There were stories of brilliant young men (this particular characterisation of gender is as we shall see both accurate and significant), inventing gadgets in garages. There was a flowering of small firms and synergy and exchange between them, though also an ideology of the free market and competition with which some of the notions of synergy and sharing the latest ideas sometimes sat a little uneasily. Big firms, sclerotic in their age and size, were seen as most important for spinning off smaller ones, like the mother-of-thousands plant, as young prodigies left to start up on their own.

There was both truth and fantasy in all this (Saxenian 1983, 1985, 1989). Both these regions in the United States were certainly focuses of extraordinary high-technology growth. And such growth elsewhere has certainly shown a remarkable tendency to cluster in particular areas, creating at least the impression of favoured ferments of new activity. But there has also been a tendency to weave general principles from what were actually more like special cases. Hewlett and Packard may have started off in a garage and ended up a multinational, but most new-firm start-ups were far more ordinary, and their development was far less spectacular. Many of the crucial companies were anyway conventionally large and multinational. Fairchild, similarly, became the model for over-generalisation, through the Fairchildren. In fact, too, defence contracts and other state expenditure, hardly the apogee of risk-taking entrepreneurial competitive capitalism, were important in both areas and have continued to be a central axis in the development of high-technology industry in many countries. Yet, as we shall see, in spite of the dislocation between representation and what a more sober assessment might point to as reality, the rhetoric was widely adopted not just by excitable journalists but also as a basis of policy.

In particular it was adopted in the United Kingdom. There are perhaps particular reasons why what we shall argue is this high-tech fantasy should have so much purchase in the UK in the current period. Since the mid-1960s but with increasing intensity over the course of three decades the UK economy and geography have been faced with the decline of some of their previously major and structuring sectors of production and employment. While some of this was interpreted as merely the passing of an historical era (an interpretation which itself, with its unimaginative division into sunset and sunrise, we shall have cause to question), it also raised issues of the longer-term decline of the British manufacturing economy. One element in the explanation of this decline, which was shared though in different guises by right and left of the political spectrum, pointed to a lack of technological dynamism within industry which itself was seen to be deeply rooted in the British class structure. The core-story of this malaise was that of the British intellectual genius who makes a major scientific breakthrough, a breakthrough with significant potential for application in practice and for commercial success. In the story, this potential remains unrealised in the UK, but is taken up and developed elsewhere, usually in the USA or in 'Europe', or more recently Japan. The UK gets the Nobel Prize and somewhere else takes the profits.

Behind this story is a whole range of shared understandings about British class structure and the cultural proclivities of particular class strata which we shall pull out and examine in the chapters which follow. Chief among them are, on the side of the more privileged strata, a disdain for making things, an anti-urbanism, and a desire to remain well removed from the actual matter of physical production, and more generally the widely acknowledged low social status of engineers and the mutual antipathy between 'real life' (whether that be the shop floor or, as it has been latterly pictured, the accounts department) and the airy-fairy world of academe. Indeed, the gap between direct production and academia has been interpreted, from the 1960s of Harold Wilson to the recent decade of Margaret Thatcher, as a crucial problem which it is essential to resolve.

In this context, the dawn of a new technological era presented an opportunity to do things differently. Moreover, as we shall see, the particular nature of the current era of high technology was especially propitious for such a move. The division of labour within industry itself was sharpened and the possibilities for geographical location were widened. Science parks became one element in a broader set of strategies designed to address the problems as perceived. From the level of national government their remit was twofold: to bring together industry and academe, and specifically to provide conditions for the more successful commercialisation of the discoveries of basic science.

It was Harold Wilson, then Prime Minister, who started the ball rolling by discussing science parks after a visit to the USA in the second half of the 1960s; and by the early 1970s the first two had begun to get under way. There was then a lull for ten years, with the real burst of activity beginning in the early 1980s. As we shall see in a later chapter (chapter 6) there were definite traces of the technocratic ideology of the Wilsonian era of white heat in the debate over some of the early development, but in the main the archetypal form of science park in the UK today reflects a thoroughly monetarist liberal ideology. This is absolutely not to say that the rationale for any individual park, put in train by a university or by a local authority for instance, lies in such a philosophy. As we shall see, especially in chapters 2, 6 and 7, the objectives of the different agents involved in their construction have varied widely (and also very interestingly, although, as we shall see, with a systematic pattern). Indeed, one or two science-park developments have been established with very different political aims in view. None the less, to do that they were wrestling with a basic form which tends in another direction and with a more characteristic development which embodies a right-wing liberal view of the world. In spite of the small scale of their actual development, science parks have come to epitomise one side of a set of distinctions. Indeed, as we shall also see, juxtaposition and counterposition are a lot of what they are about. They represent sunrise as opposed to sunset parts of the economy. They stand for small firms and individual genius rather than dull corporations, and for individual genius commercialised rather than doddering ineffectually in some university laboratory. They represent workers (or, rather, employees and small entrepreneurs) fulfilling themselves through commitment to their work. They exemplify the possibilities of new forms of work organisation. They epitomise how much better everything can be (it is said) without trade unions. In brief, they are everything which nineteenth and early twentieth-century manufacturing is not.

That, of course, is the rhetoric, and one thing we shall be doing (in chapters 2, 4, 6 and 7, for instance) is investigating how far it is fulfilled. But, however far it is fulfilled, the rhetoric has its own effects, and science parks have become powerful symbols of a new approach to issues of industrial regeneration.

Clearly, the current ideological gloss to science parks is not 'neutral', as attempts to work it otherwise demonstrate. But in this book we want to argue more than that. The argument is that the underlying founding premisses on which science parks and a whole range of related technological and industrial strategies are based are also, though much more rarely questioned, in fact profoundly problematical. Fundamental to this argument is the conceptualisation of science parks as an object of study. In this book we work with two conceptualisations of science parks. First, there is what might be called the popular conceptualisation, the one which is used, for instance, in the policy literature produced by the parks themselves and by journalistic commentaries, as well as by some academic literature which takes those concepts as given. This conceptualisation is important; partly, as we have said, because such formulations, whether accurate or not, produce their own effects and partly because it is within this formulation that science parks as a policy of economic regeneration are most frequently judged. We, too, begin by assessing science parks 'in their own terms'. But we also assess the popular conceptualisation itself, and find it lacking. Second, therefore, we present our own alternative way of conceptualising science parks. This is developed step by step through the book but, briefly, it interprets science parks as, in their archetypal form, composed of three aspects: first, they are based on a particular model of scientific investigation and industrial innovation; second, they have a particular spatial form and content; third, they are property developments carried out by particular agents with specific interests. Each of these aspects is developed further in the chapters which follow. Moreover, it is argued that each element of this definition has necessary consequences which are quite different from those postulated in the popular model. Further, we would argue, those consequences are in the end quite detrimental to the objectives which science parks are established to achieve. Moreover, these real consequences are, we argue, certainly detrimental to any progressive politics. Not as a mere accidental outcome, but as an implication of their very conceptual foundations, policies such as science parks lead, we argue, to increasing social polarisation and growing geographical inequality.

One theme, then, revolves around issues of conceptualisation. It will weave itself throughout the argument of the book. Moreover, it involves the active articulation together of social, spatial and scientific/technological characteristics and a careful distinction between the necessary implications of the archetypal model of a science park and the consequences of the contingent circumstances of their current development.

A further theme focuses particularly on the geographical aspects of these developments, and in a number of ways. Science parks are often promoted as policy instruments which will have an impact on the geographical distribution of industry and employment. One of our arguments is that in their current form this impact is marginal and likely to remain so. What emerges, however, is that science parks take radically different forms, in terms of their relation to the local economy, the kinds of companies located in them, their political and financial purposes, and even the relation between public sector and private. And these different forms have a systematic geography, distinguishing the north of the country from the south, and inner-city areas from the smaller towns of the semi-rural sunbelt. More even than this, however, we argue that spatial form and spatial symbolism are integral to the archetypal science park and that these characteristics interact with the social and technological ones with potentially, and most usually, negative results.

Finally, there is an argument about science and technology, which again has a number of threads. Science parks are based on a highly particular model of the processes of scientific discovery and technological innovation. Our argument is that, not only is this model not the only one, it is also inaccurate as a description, including of what science parks in fact do, and potentially problematical. This is so because of the relation between that model of innovation and the divisions of labour which it implies, and structures of social inequality. Further, we argue, not only does this model of the science–industry relation lead to particularly inegalitarian social structures but, combined with the spatial content of the archetypal science park, it can only serve to reinforce geographical uneven development. Finally, to complete the circle, evidence from both this country and elsewhere indicates that both social structure and geographical form are likely to have – exactly contrary to most of the literature on the subject – a detrimental effect back on the processes of technological advance and economic regeneration.

The arguments of the book begin in earnest with chapter 3, which explores the assumptions about the nature of scientific investigation and industrial innovation which, we argue, are implicit in the archetypal notion of a science park. This is the first axis of our reconceptualisation, and the chapter explores the particular history of this model of science and innovation within the United Kingdom. It also begins to explore some of the social and spatial implications of this model and to show how it is increasingly subject to challenge.

One of the implications of the science-park model of innovation is a hierarchical division of labour and a differentiation of status between the various parts of that division of labour. Chapter 4 picks up this issue by investigating the nature of the labour process in establishments on science parks. In particular, attention is paid to issues of flexibility and of control over the work process. What emerges from this empirical investigation confirms the arguments of chapter 3: that the undeniably good conditions for the few are not generalisable to the majority. Science parks cannot be held up as a vision of the future of employment for all members of the workforce.

Chapters 3 and 4 thus both point to issues of social position and status and chapter 5 takes these up directly. It examines the class character of scientists and technologists, and the basis of their social standing. It is argued that, certainly in the British case, this basis (at least rhetorically) has changed considerably in the last decade. Moreover the current status of scientists and technologists is further shored up by other factors, among them the fact that in the UK they are in national shortage and – perhaps less well known – their geography. Here the second aspect of the definition of science parks – their spatial form – is developed, and it is shown how that too plays an active role both in the construction of social status and in the effectivity of industrial policy.

Chapter 6 examines the issue of geography from a different angle and explores science parks as part of local economic strategies, taking one case from a high-tech growth area and a second from an inner city. While the conclusion must be that, overall, the potential of science parks as regenerators of local economies is limited, perhaps the more important implication of this very different form of 'policy evaluation' is that there are – building on the arguments of previous chapters – contradictory tensions at the heart of the strategy itself.

These issues show up in distinct ways, however, in the two areas studied in detail, so chapter 7 steps back to examine the relation between public sector and private across the country as a whole. Here we build on the third axis of the reconceptualisation of science parks – as property developments. What emerges is that while there is everywhere a relation between public sector and private, the nature of it varies across the country, and is in many ways perverse. In areas which are already prospering – especially in the south and east – science parks are profitable for the private sector on the back of already existing growth; in the 'north' the public sector struggles to provide some counter-balance; in between, 'partnerships' between the two sectors can all too easily lead to the public subsidising the private while the latter undermines the former's initial objectives.

All these debates have led a long way from what is normally called policy evaluation, and what they all imply is a deeper rethinking of science/industry policy, of regional policy and of local economic policy than such evaluation would usually indicate. We begin the book, however, with the more usual form of evaluation. Chapter 2 takes science parks in their own terms: an examination of science parks as they are most usually conceptualised and an assessment of them in terms of the aims and objectives most frequently stated for them. In this way chapter 2 also enables us to lay out a considerable amount of factual information about science parks as they are in the United Kingdom today, which is important for the arguments in the rest of the book.

2

SCIENCE PARKS IN THEIR OWN TERMS

THE POPULAR CONCEPTUALISATION

There is a popular conception of science parks, one which is used in the policy literature produced by the parks themselves, and reproduced by journalists and by other writers. It comes in two parts. First, there is the actual definition of what a science park is, and second there is a set of (often implicitly) postulated causal relations, effects which it is assumed will spring from these characteristics.

Definition, and a thumbnail sketch

The number of science parks in the UK depends on how they are defined, even within the popular conception. Numerous 'high-tech' or otherwise 'up-market' industrial estates exist, with a bewildering array of titles, from 'technopark' to 'business park' to 'enterprise park'. One 1983 report identified sixty science-park-type schemes (Debenham, Tewson & Chinnocks 1983), and Henneberry counted 101 'high-tech property schemes' existing or proposed up to October 1983 (Henneberry 1984a). The examples at the beginning of chapter 1 cover quite a variety of phenomena. In fact, however, not all these different developments are really science parks, for, fortunately for our purposes, there has been explicit consideration of what does and does not qualify for membership of this quite carefully defined group. In the United Kingdom this definition has been produced by the UK Science Park Association (UKSPA).

When UKSPA began in 1984 it published a set of criteria for membership eligibility:

A science park is a property-based initiative which:

- has formal operational links with a university or other higher educational or research institution (HEI)
- is designed to encourage the formation and growth of knowledge-based businesses and other organisations normally resident on site
- has a management function which is actively engaged in the transfer of technology and business skills to the organisations on site

(UKSPA 1985)

It is parks which fulfil these criteria which form the focus of attention in this book. For parks which accord with this definition form a coherent group with underlying conceptual and causal similarities (although we shall be questioning the *nature* of these as postulated in the model laid out in this chapter). The book is therefore not concerned with the plethora of high-technology industrial estates which have also sprouted up in recent years. For these are not founded on the same causal logic, nor do they embody the same underlying principles, as do science parks.

According to UKSPA, there were thirty-eight parks fulfilling the criteria for qualifying as a science park in operation at the end of 1988, with one more under construction (see figure 2.1), and a further eighteen parks were at the planning or feasibility-study stage (UKSPA 1989). These data exclude parks such as Wavertree Technology Park in Liverpool, where the Plessey company has been involved as host institution, and Birchwood Science Park at Warrington, which has some links with neighbouring British Nuclear Fuels Ltd. These parks were excluded from the UKSPA listings because of the lack of a formal link with an academic establishment. However, UKSPA has now shifted its academic-link criterion to include not only university and other higher educational research institutions but also major centres of research. Thus the new Billingham park, with its strong links to ICI, is now eligible to join UKSPA.

Science parks are a relatively new phenomenon. Table 2.1 shows when UK parks opened. As noted in chapter 1, the most obvious feature of this chronology is unevenness over time: two parks opened in 1972, there were no more for ten years, and then followed the main growth period in the 1980s.[1]

Figure 2.1 Science parks eligible to join UKSPA (end 1988)

Table 2.1 Science parks by year of opening

Year	*Science park*	*Number of parks started in year*
1972	Cambridge, Heriot-Watt	2
1982	Merseyside	1
1983	Aston, Bradford, Leeds, Glasgow	4
1984	East Anglia, Hull, Loughborough, Manchester, Nottingham, Surrey, St Andrews, Southampton, Warwick	9
1985	Aberystwyth, Clwyd, Durham, South Bank, Sussex	5
1986	Antrim, Birmingham, Bolton, Brunel, Kent, Stirling, Swansea	7
1987	Cardiff, Keele, St Johns (Cambridge), Bangor	4
1988	Sheffield, Aberdeen, Sunderland, Billingham, Salford, Wrexham	6
	Total	38

Source: UKSPA listings

It is also apparent that there is a wide geographical spread of science parks across the country (figure 2.1). What is more, there is a strong North–South divide. Table 2.2 gives the first indications of this. On the one hand there are, perhaps in itself surprisingly, more science parks in the north of the country than in the south, but on the other hand the parks in the south are more developed. While the south has 24 per cent of parks, its proportions of build-

Table 2.2 The North–South divide in science-park development (percentages in brackets)

	Number of parks	*Area of buildings (000 ft²)*	*Number of tenants*	*Buildings: area under construction (000 ft²)*	*Employment*
South	9 (24)	1,360 (43)	259 (32)	659 (55)	5,178 (49)
North	29 (76)	1,787 (57)	548 (68)	543 (45)	5,362 (51)
Total	38	3,147	807	1,202	10,540

Note: The North–South divide here is demarcated by a line running from Bristol to the Wash. This is, in general, the 'North–South' divide we have used throughout the book. Because of its extension into East Anglia, especially Cambridge, the 'south' is sometimes referred to as the 'south and east' of the country. Except for in the context of the North–South divide, the use of capitals for region names indicates a reference to a Standard Region

Source: UKSPA (1989) Summary of Operational Science Parks in the UK

ing area, firms, buildings under construction and employment are all higher than this. No serious conclusions can be drawn from these figures – in part they reflect the fact that some of the parks in the north are still extremely new. None the less, the figures do point to an important line of enquiry – the dimensions of a North–South divide within the UK science-park phenomenon will remain a theme throughout this book. One small pointer to one of the differences between north and south which we shall examine later can already be picked up from these figures. This is that, in terms of both employment and physical area, the north would seem to have a higher representation of smaller firms than the south (it has 68 per cent of the total firms but only 51 per cent of the employment and 57 per cent of the building area).

All such averages have to be treated with caution, however, for there is wide variation between individual parks and individual firms. Table 2.3, which ranks UK parks in terms of their number of tenant establishments, gives some indication of this. The ten parks with the most tenants have over half of all establishments whilst the eleven parks at the bottom of the list, each with ten or fewer tenants in 1988, account for less than one-tenth of establishments. Moreover, the rate of increase in numbers of firms also varies greatly, witness the high growth of Surrey and Brunel in only two years. Interestingly the two oldest parks show different growth-paths. Heriot-Watt took sixteen years to reach that number of occupants. Cambridge, the largest science park in the UK, grew slowly from its 1972 beginnings, with only seven tenants in 1978, but then the number of tenants grew rapidly in the early 1980s, only to slow down again subsequently. Once again, it is important to be careful about drawing conclusions from such figures. In the case of Cambridge this latest slowdown certainly does not necessarily indicate any shortage of demand. Rather it may indicate a shortage of building-space on the park. The kinds of data frequently used to judge the success or failure of science parks often raise as many questions as they answer.[2]

But if we pursue further the question of investment more issues can be drawn out. By late 1990, £293 million had been invested, in total, in land and buildings in UK science parks. This investment came from a variety of sources (see table 2.4) with, perhaps against popular expectations, 59 per cent of the total coming from the public sector (local authorities, academic institutions and

Table 2.3 Science-park tenants, 1986–8

Science park	*Number of tenants* *1986*	*1988*	*Percentage change 1986–8*
Cambridge	68	69	1
Aston	42	60	43
South Bank	36	50	39
Warwick	35	47	34
Surrey	10	43	330
Bradford	26	33	27
Heriot-Watt	23	32	39
Sheffield	–	32	–
St Johns (Camb.)	–	31	–
Billingham	–	30	–
Brunel	8	25	213
Durham	5	25	400
Nottingham	14	22	57
Glasgow	15	21	40
Southampton	15	21	40
Swansea	6	20	233
Clwyd	6	20	233
Salford	–	18	–
Manchester	11	17	55
Stirling	5	17	240
Bolton	10	16	60
Loughborough	17	16	-6
Merseyside	12	15	25
Cardiff	–	15	–
Birmingham	9	15	67
Hull	12	14	17
Sussex	–	14	–
Aberystwyth	6	10	67
Keele	–	10	–
Sunderland	–	9	–
Leeds	11	8	-27
Antrim	2	6	200
Aberdeen	–	6	–
Wrexham	–	5	–
Kent	2	5	150
Bangor	1	5	400
St Andrews	2	4	100
East Anglia	3	1	-67

Source: UKSPA

Table 2.4 Sources of investment in UK science parks (percentage of total expenditure)

Source	*1988*	*1990*
Academic institution	28	29
Local authorities	11	9
Development agencies	21	21
Private-sector finance	8	16
Tenant companies	32	24
Other	-	1
Total	100	100

Source: Interviews, and UKSPA

development agencies). There is, moreover, a strong regional variation in public–private investment. Apart from the contribution of host academic institutions, which for the most part takes the form of land rather than direct financial investment, in the 'southern sunbelt' only two parks out of nine, Southampton and Kent, had public-sector investment. Outside the sunbelt only one park out of twenty-nine, Heriot-Watt, did not have direct public-sector investment (interview). Even here, the Scottish Development Agency (SDA) later funded an incubator unit. The level of property investment also varies considerably between parks. The Surrey Research Park has buildings financed by Grand Metropolitan and BOC at a cost of £20 million, with Surrey University funding its proportion of initial investment by the sale of land to these corporations. Warwick attracted some initial private investment – from Barclays Bank for an incubator unit. Public funding stepped in to fund 'high risk' larger premises – West Midlands County Council provided £1 million and Coventry City Council a further £1.2 million. This contrasts with the Merseyside Innovation Centre where a former Liverpool University physics building was refurbished with grant-aid from the Inner City Partnership Scheme.

Finally, in this thumbnail sketch, what about employment? Table 2.5 gives basic information but, as with everything else, averages are virtually meaningless. In the case of employment the picture is dominated by the larger parks and by a handful of large establishments. Table 2.6 gives an indication of employment by park, and once again there is a strong difference between north and south. As table 2.2 indicated, the 24 per cent of parks in the south have 49 per cent of the total employment. Moreover our data confirm the difference in size of firm between north and south of the country. The average employment per tenant in the south is 20.0; in the rest of the UK it is 9.8.

Table 2.5 Park tenants and employment, 1985–90

	1985	*1986*	*1987*	*1988*	*1990*
No. of parks	21	28	33	38	39
No. of tenants	301	412	642	807	1,012
Total No. employed	3,800	5,300	7,600	10,540	14,708
Average employment per tenant	12.6	12.9	11.8	13.1	14.5

Source: UKSPA

Postulated effects

If this, then, is the UKSPA definition of science parks, and something of what they are like, what is it supposed will be their effects? The best way of addressing this question is through an analysis of policy statements, press cuttings and papers produced

Table 2.6 Employment in science parks (estimates)

Science park	*1988*	*1990*
Cambridge	2,500	2,856
Surrey	1,500	2,000
Aston	735	1,000
Warwick	700	999
Heriot-Watt	610	700
South Bank	457	864
Wrexham	350	496
Bradford	300	410
Hull	248	162
Nottingham	240	466
Sussex	240	200
Southampton	225	282
Billingham	200	300
Other	2,235	3,973
Total	10,540	14,708

Source: UKSPA (1989, 1990)

by those setting up science parks in the UK in the 1980s. This revealed over twenty-five different aims of parks as stated by their management and sponsors. Table 2.7 lists these aims.

Table 2.7 Summary of science-park objectives

To:

Stimulate the formation of start-up new-technology-based firms (NTBFs)
Encourage the growth of existing NTBFs
Encourage spin-off firms started by academics
Encourage and facilitate links between higher education institutes (HEI) and industry
Facilitate technology transfer from academic institution to park firms
Commercialise academic research
Increase the 'relevance' to industry of HEI research
Give academic institutions access to leading-edge R&D in park firms
Increase the appreciation by academics of industry's needs
Create employment and consultancy opportunities for academic staff and students
Foster the technologies of the future
Create synergy between firms
Create new jobs for the region
Improve the performance of the local economy
Improve the image of the location, particularly for areas of industrial decline
Shift local/regional industrial base from declining to new industries
Counter the regional imbalance of R&D capability, investment, innovation
Stimulate a shift in perceptions
Build confidence
Engender an entrepreneurial culture by example
Reproduce Silicon Valley, Hewlett Packard
Attract inward investment, mobile R&D
Provide an adequate and safe return on capital
Generate income for the academic institution
Stimulate science-based technological innovation
Improve the image of the academic institution in the eyes of central government

In some cases the aims were at the outset less than clear. For example, Ian Page, Economic Co-ordinator of Bradford City Council, reflected to the UKSPA Conference in 1985 that the council had very little idea what they were getting into when they decided to have a science park in Bradford. There were 'No in-depth studies, no academic research, just an idea. Let's have a Science Park in Bradford' (Page 1985: 1). Nevertheless, Page emphasised that there was an underlying aim for the council in promoting the park: to change the image of Bradford away from declining engineering and textile industries and derelict factories. Associated with this aim was the practical one of raising industrial rental levels in the city – 'It is not good to have the lowest rents in the country' (Page, spoken comment, UKSPA Conference, 1985).

The Bradford Science Park was financed by English Estates, for whom it was their first 'high-tech' development. English Estates' objectives were:

> a. To encourage interaction between industry and university in Bradford, the transfer of technology and the development of new marketable products and services
> b. to promote Bradford as a centre for knowledge-based companies
> c. to provide good quality premises for established or start-up companies engaged in high-technology activities.
>
> (Pender 1985: 2)

For the university, a major objective for the park was to 'create jobs for local people and graduates in a suitable environment close to the campus. After the 1981 cuts it became clear that we would not need land earmarked for future expansion. We decided to support its use for the formation of companies that would interact with the university' (interview).

Here we find some of the recurring themes surrounding science-park initiatives: university–industry linkage and technology transfer, the promotion of new start-up companies, the focus on 'high' technology, and the creation of new jobs.

These themes were reinforced by Lord Young, then Secretary of State for Employment, in his opening address to the 1985 UKSPA Annual Conference:

> Science parks have much to do with the wealth and job creation that comes from enterprise, small firms and new technology ... I strongly believe that the education system has a vital role to play in achieving this, and science parks are precisely about these issues ... I believe that one of the long standing problems in this country was the separation of the 'groves of academe' from industry and from wealth creation ... It is important that industry and education are linked more closely together ... There is still a long way to

> go but the active involvement of academic institutions in providing not only property but ideas and support for new technology based firms is an indication of the way we should be developing for the future.

As a further example, Jeffe Jeffers, Project Director and 'driving force behind the South Bank Technopark' in London (*Technopark Link* Newsletter) stated that the principal objective of the Technopark was 'to keep good quality industry in the inner city' (Marsh 1984). Marsh summarised the implications of this: 'Companies that start off in the park will create employment in the area through fast growth'. Indeed, an even more ambitious goal for the Technopark was envisaged by the Pro-Assistant Director of South Bank Polytechnic: 'There could be a Science City at the Elephant and Castle, a chip valley right in the middle of a derelict area of the city is poo-poohed by a lot of people but this could be the beginning of that' (interview with Christopher Price, former Labour MP, Pro-Assistant Director of South Bank Polytechnic, *The Times*, 9 December 1985).

The two themes of stimulating high technology and the promotion of university–industry links were continued by Dr Malcolm Parry, Marketing Director, Surrey Research Park, who identified the 'very clear aims' of the park there: 'to develop income from industrial sources for the university, help further the strong links with industry already enjoyed by the university and develop opportunities for people in the university to engage in consultancy work' (interview in *Financial Times*, 30 November 1985: xii). By 1988 the park was indeed generating 'both direct income and the growth of a capital asset' (Malcolm Parry, personal communication). The university, through its 100 per cent ownership of the park, owned all except BOC's building, having bought back the Grand Metropolitan incubator unit for £1 million.

Support for new firm formation was emphasised by the Director of the Merseyside Innovation Centre (MIC):

> MIC was set up in 1982 to assist in the regeneration of the economy of Merseyside in defined, specific ways ... We encourage and organise transfer of technology between the University and the Polytechnic and industry and commerce on Merseyside ... We are involved in the creation of new ventures based on technological developments from whatever source ... we offer nursery accommodation and common services ... we

> need to offer advisory and information services to an ever-increasing clientele ... We are a small part of the education and training network on Merseyside.
>
> (Rimmer 1985: 1)

Cambridge Science Park (CSP), on the other hand, has not felt the need to prioritise industrial regeneration. Neither has it felt the necessity for a substantial 'hands-on' management team. The Senior Bursar of Trinity, Dr John Bradfield, has been involved in the park since before its inception, writing papers on the subject as long ago as 1971 (Carter and Watts 1984). He defined the purpose of the CSP in 1983 to be to:

> meet government requests for greater interchange of ideas and people – and greater sharing of equipment and libraries – between universities and high technology industry. Close proximity between the two is desirable for a variety of inter-related and obvious reasons.
>
> (Bradfield 1983: 1)

These include technology transfer, often through informal mechanisms, commercial feedback to the university, assisting the awareness of industry's problems and aspirations, and also stimulating the commercial exploitation of university potential. The park would generate prosperity for the nation by the commercial development of new scientific ideas. Bradfield also stated that the CSP facilitated qualified staff recruitment for companies sited there, and widened 'job opportunities for graduate and non-graduate staff'. Students would benefit from familiarisation with high-technology industry, 'which is helpful to them in formulating their employment aims' (Bradfield 1983). Bradfield does not mention the formation of start-up companies as such, but writes more generally about the 'potential in the fountain of new scientific ideas which flows in a great university'.

The example of Cambridge Science Park also again illustrates one aim of some science-park projects: to obtain a return on investment in the park. According to Carter and Watts:

> The development and management objectives reflect the college's dual role: first in the property context as land-owner, funder and developer; and secondly, as a leading academic body. The prime objectives were to achieve a commercial investment return commensurate with prime

> industrial property and successful development which would not be a drain on Trinity's portfolio ... Other objectives were to utilise unproductive land whilst at the same time removing an eyesore in the proposed green belt; to encourage the development of science based industry as envisaged in the Mott report; and, to provide the flexibility to meet the rapidly changing needs and demands of innovative science-based firms.
>
> (Carter and Watts 1984: 21)

Many of those interviewed for this study, including the directors of tenant firms, perceived recent changes in Trinity's strategy, and particularly in its interest in fulfilling a management role beyond property management. Only recently had Trinity begun actively to encourage formal or semi-formal links between the university and the park tenants. One senior Cambridge academic, who was at the time also a director of a science-park company, took the view that 'Cambridge Science Park has evolved backwards – it started with no support for firms, and has only recently begun to think about that' (interview).

What seems to emerge from the development of Cambridge Science Park is what Segal Quince describe as 'an unhurried long-term approach, working informally and in line with the natural grain of events ... It is perhaps a "laid-back" approach compared with arrangements elsewhere, but it is realistic and it is working' (Segal Quince 1985: 41). However, many observers have commented that CSP has been able to follow rather than lead the market for 'high-tech' property in Cambridge. It does seem that almost any property initiative would stand a good chance of success in filling its accommodation in Cambridge.

Cambridge Science Park's 'hands-off' strategy is in marked contrast to Aston Science Park's approach in Birmingham, where there has been a strong proactive management strategy, coupled with a series of statements of aims from those groups involved in establishing the park.

Aston Science Park involves co-operation between Aston University, Lloyds Bank and Birmingham City Council. For the council, the key aim has been 'regeneration ... a determination to see the economic resurgence of Birmingham' (Aston Science Park 1986: 3). The aims and priorities have varied a little, depending on which was the controlling group – with Conservatives stressing the creation of 'wealth' and new businesses, and the Labour group emphasising the regeneration of industry and the creation of jobs. For Lloyds Bank its 'roots as a bank grew out of the beginnings of Birmingham's industrial development in the eighteenth century' (quote from Sir Jeremy Morse, Chairman, Lloyds Bank, to Cookson 1982). It contributed £1 million to the park's venture capital fund, an acceptance of risk in expectation of long-term return. The university has emphasised the interaction between 'the academic world and that of industry and commerce' (Aston Science Park 1986: 2), and has seen the park as a way of improving its own standing.

Aston VC Sir Frederick Crawford (credited with the idea of the park) saw the venture early on as a 'powerful wealth and job creating centre'. 'The industrial park at Stanford has grown into Silicon Valley. We are trying to reproduce some of that ferment here both for the university and for industry' (Rogers 1985). Early statements from Birmingham councillors, and the press, emphasised the potential of new technology and its job-creating abilities. Councillor Wilkinson argued that the park represented 'a major initiative to shift the economic base of the city towards the enterprises and industries of the future' and 'forecast that within five years, 5,000 new jobs would be created directly on an expanded science park and another 10,000 generated indirectly' (Cookson 1982: 11). The *Financial Times* reported that 'It is hoped the scheme will create up to 15,000 jobs in the West Midlands ...' (*Financial Times*, 24 November 1981: 8).

A succinct account of the university's view of the intermediate and long-term objectives of Aston Science Park was provided by Professor Foster, Pro Vice Chancellor and Professor of Mechanical Engineering at Aston University:

> The objectives of the Park are to encourage the formation and growth of a wide range of new, small and technologically oriented companies, and through the links that these companies will develop with the City and the University, to improve employment prospects in Birmingham and improve the economic strength of the West Midlands.
>
> (Foster 1982: 11)

Aston has set up a nine-strong management team working for Birmingham Technology Ltd (BTL), a company set up by the three sponsoring organisations. It has evolved a dual strategy of proactive management, which includes support for new small firms, and flexible premises. It is able to provide larger accommodation to firms as they grow. BTL provides equity capital to many firms, a facility unique to Aston Science Park in the mid-1980s and a major policy tool enabling Aston, perhaps more than any other park, to put the policy objective of encouraging start-ups into practice. BTL Chief Executive Harry Nicholls emphasises the long-term nature of the Aston Park strategy: 'Science parks are not a "quick fix"' (Nicholls, spoken comment, UKSPA Conference, 1985). The growth of companies from start-up takes time and requires much support from BTL management.

Among UK science parks, Aston is undoubtedly more typical than Cambridge in having to 'create the market' for the science park itself, in an inner-city area with little existing demand. Moreover, while Aston has gone further down the proactive route than have others, its aims find echoes in many newer parks. Thus a manager of one northern science park mentioned that the aim of the city council was to create employment and encourage high-technology industry in the city, whilst improving the local environment. Or again, for Manchester University the aim was to encourage the take-up of technology in industry, and also to generate income. 'The science park in Manchester is an integral part of the university's increased commitment in technology transfer.' The science park is a 'strictly commercial operation' (Richmond 1985). But it too has a number of different sponsors – the city council, the university and four companies with major local connections.

The aims of two other northern parks were agreed by managers in 1986 as being:

(i) To satisfy the (political) need for universities to be seen to be having contact with industry.
(ii) To help academics put their ideas to commercial use.
(iii) To help companies which need the expertise provided by the university.
(iv) To stimulate employment generally in less 'well-off' areas.
(v) To improve technology.

The first manager of the West of Scotland Science Park summed up his park's aims as being:

> to reinforce and expand the technology base of Scottish industry by: facilitating the profitable transfer of academic knowledge and developments into new and existing

businesses, attracting R&D-based business units into a supportive location and encouraging successful new business start-ups in high technology areas.

Additional objectives are those of commercial return on the development and contribution to other initiatives of the three partners in the project.

(Bond, in Gibb 1985: 136)

A subsequent manager distinguished between the objectives of the Scottish Development Agency (SDA), which funded the park, and those of the two universities involved, Glasgow and Strathclyde. The objectives of the SDA are to encourage technology transfer via (i) academics commercialising their own ideas, and (ii) existing businesses utilising university facilities and expertise. The universities' objectives were (i) to improve industry's use of university expertise and R&D, and (ii) to generate income in order to supplement UGC funding (interview).

This has been a reasonably representative journey around a significant group of UK science parks (and it draws on a much wider survey and analysis). It has allowed us to flesh out the aims of the science-park movement. Certain dominant reasons for establishing science parks constantly recur. These are that:

- (i) Parks will promote the formation of new firms.
- (ii) They will facilitate links between the host academic institutions and park firms and thus improve the take-up of ideas to new products and processes.
- (iii) Firms on the parks will have a high level of technology and be 'at the leading edge'; they promise a sunrise future, in many areas to replace a 'sunset' existing local economy.
- (iv) They will create employment opportunities.

These, then, are the assumed potential effects of setting up a science park. They imply that if a park is established in conformity with the concept defined at the beginning of this section, then certain things will happen, of which these four emerge as the most important. There is an assumption, in other words, about causal relations. It is these which we shall now investigate.

The four aims together amount in many cases to attempts to revitalise local economies, and chapter 6 of this book will consider in depth that wider aim. But as that chapter and chapter

7 will explore, and indeed as this chapter has already hinted, the aims of science parks vary depending on the part of the country in which they are located. A science park in an inner city or a northern manufacturing town probably has different aims (as well as having different potential, and different obstacles in its path) from a science park in the sunbelt of the south. There may be contrasts, too, in the objectives of the various sponsors of parks – a very different objective from the four above, but one referred to by a number of the people we have quoted in this chapter, is to provide a commercial return on capital investment. The conflicts which may arise between objectives, and their relation to the geography of park aims, will be explored in chapter 7.

But what we shall do in the rest of this chapter is focus on the four aims (postulated effects) highlighted above. They provide specific criteria, from the statements of those responsible for setting up and running science parks in the UK, against which parks may be evaluated. Others have attempted such an exercise, and have usually emerged with a pronouncement either that science parks are a runaway success, or that they are a failure, a drop in the ocean. Our assessment is more measured. For one thing, science parks so far have only fairly short histories. Yet many of their aims are necessarily long term. Indeed, there is a tendency for more recent statements of objectives to be more circumspect and to stress rather more strongly the long-term nature of their horizons. Yet all the aims we have highlighted remain.

It is therefore on these that the rest of this chapter will concentrate. Effectively, we are evaluating science parks on their own terms, based on their own self-conceptualisation and on their own stated objectives.

The data we draw on come from six sources. First, the United Kingdom Science Park Association (UKSPA) has annually, since 1985, published a set of basic data on science parks. These include number of parks, tenant companies, building area, parks planned for the future, and so on. We have used all these data sets for our chronological tables, but primarily the data obtained for end-1988 (UKSPA 1989). Second, UKSPA published a *Tenants Directory* in 1987, with data on 224 tenant establishments of the around 400 then existing (UKSPA 1987). Third, in mid-1986 we conducted, with UKSPA and the University of Newcastle Centre for Urban and Regional Studies (CURDS), a survey of 183 establishments on twenty UK science parks, together with a comparison group of 101 establishments located off-park. The 183 park units made up over half the total number of establishments at that time. (Establishment numbers rose from 300 to 391 during 1986.) The off-park group was selected as a control, to be of similar age, size and sector.[3] The survey was sponsored by Peat Marwick McLintock. It is referred to here as the UKSPA-OU-CURDS survey. Fourth, as part of our two-year project funded by the Joint Committee of the Economic and Social Research Council and the Science and Engineering Research Council, 'Science parks and industrial innovation in Britain', we have a data base on more than 200 science-park establishments. Fifth, for the same project we conducted a series of eighty-one in-depth interviews with nine park managers, forty-eight establishment managers and employees, ten local economic and other agencies, and fourteen key people involved in the setting up and running of science parks. Interviews focused on Cambridge Science Park and Aston Science Park but were not confined to them. Sixth, we were able to draw on a follow-up survey to UKSPA-OU-CURDS carried out in late 1990 (Storey and Strange 1990).

EVALUATING SCIENCE PARKS ON THEIR OWN TERMS

The formation of new start-up firms

That parks will facilitate the establishment of new firms emerged as a strong theme in our review. The perceived importance of the entrepreneurial start-up firm in the US science-park model, linked with such names as Hewlett Packard and Apple Computer, is reflected in the UK in the aim of being a base for fast-growing and independent start-up companies. So are firms on science parks predominantly new independent start-ups or are they relocations of existing businesses or branches of larger companies?

A range of information is available to show that new independent start-up enterprises do establish themselves on science parks but that they are in a minority. UKSPA data in 1985 suggested that, out of 300 establishments on science parks, 70 per cent were relocated units or firms and 30 per cent were 'new starts' (Dalton 1985). More comprehensive data from the 1986 UKSPA-OU-CURDS survey of twenty science parks, together with further interview data, revealed that two-thirds (119 out of 185) of science-park establishments surveyed had been previously located elsewhere (table 2.8). Forty-nine out of 185, a little over a quarter, were independent start-up companies.

Table 2.8 How many park units are new start-ups? (percentages in brackets)

Status		*UKSPA-OU-CURDS survey, 1986, plus other interviews*		*UKSPA Tenants* Directory, 1987*	
New	Independent	49	(27)	58	(29)
	Non-independent	17	(9)	16	(8)
Relocation	Independent	86	(47)	74	(37)
	Non-independent	33	(18)	52	(26)
Total		185		200	

* We have categorised establishments that renamed themselves on moving as relocations

The January 1987 UKSPA *Tenants Directory* (again a sample) corroborates these data (table 2.8). Of 200 establishments for which there is relevant information, 126 (almost two-thirds) are relocated establishments. Of the seventy-four new establishments sixteen are known to be subsidiaries or branches of parent companies elsewhere. Thus a maximum of fifty-eight (29 per cent) are independent start-up companies. Twenty-five to 30 per cent is not insignificant as a proportion for start-ups, even if it does not wholly corroborate the popular image of science parks. A number of points should be noted, however. First, our results contrast with claims by others. Thus one widely-quoted private consultant's report produced a figure of 47 per cent for new start-ups. Concerned by the disparity between this figure and our own, we went through the consultant's data with care. We found errors in classification, in the cases we were able to check in detail, of 30 per cent. They consisted of 25 per cent of firms classified as being new when in fact we know they were not, and occasional firms (5 per cent) being classified as independent when we know they were subsidiaries of other companies. This indicates a reduction of the 47 per cent to around 33 per cent, and the real figure could be even lower, since we were relying on 'negative evidence' rather than doing a positive check on the legal status and date of formation of each establishment. These corrections bring the figure much closer to our own.

However, even this figure has to be modified by recognition of the varied nature of the start-ups. Thus, of the fifty-eight independents in the *Tenants Directory* in 1987 a fair few were set up by park managements and public enterprises for business support and social reasons. Such different forms include Business in Liverpool Ltd, MIC Computing Services Ltd, MIC Biosciences Ltd, MIC Medicall Ltd (all on Merseyside Innovation Centre), Contact, and Manchester Industrial Liaison Centre, both on Manchester Science Park, New Tech on the Newtech (Clwyd) Science Park, and the Women and Work Training and Resource Centre on Aston Science Park.

There is also variation between science parks in the importance of new start-ups. Our own further research suggests that Cambridge, the largest park, and perhaps ironically in this context the source of most science-park imagery in the UK, has a low level of new firm formation, while the park with the highest percentage of new firms is Aston. As mentioned above, Aston has a strong policy of promoting start-up firms with its own venture capital fund (funded by the city council and Lloyds Bank). On Cambridge Science Park new start-ups are the exception. Because our own detailed survey was of only a sample of firms, our results were cross-checked with the *Cambridge Science Park Directory*. The directory reveals that thirteen out of sixty-three establishments present in mid-1986 (21 per cent) were independent firms with no previous address (i.e. slightly below an average itself heavily influenced by Cambridge). By January 1990 only eleven out of eighty park establishments were independent with no previous address. In contrast, similar information for Aston shows that eighteen of thirty-two (56 per cent) were independents with no previous address. A number of other parks have, like Aston, a high proportion of new establishments. These include Bradford, Merseyside, and Nottingham (UKSPA 1987).

Our conclusion must be, then, that science parks have been only moderately successful in their stated aim of specialising in being seed-bed areas for the start-up of new independent companies, at least in the sense that it is other kinds of capital which are taking most advantage of their existence.

On the other hand, this does give the positive result that science-park firms have an impressively low rate of failure, at around 2.5 per cent per year. In part this of course reflects the fact that the firms are, in the main, not new starts but rather relocated firms with a track record. Only a relatively small percentage of firms are less than four years old – 28 per cent according to the UKSPA-OU-CURDS survey (table 2.9). It also reflects the selection process by which park managements allow on to the park only those firms likely to succeed. The 1990 follow-up survey of the 183 establishments interviewed in 1986 (UKSPA-OU-CURDS) gives further evidence (Storey and Strange 1990). This showed that over the four-year period a minimum of fifteen and a maximum of thirty-one establishments had failed. This is certainly not a high rate, though the (favourable) comparison frequently made with UK small firms is illegitimate, since by no means all science-park establishments, as we have seen, are small independents. Indeed, the same follow-up survey also shows that, among survivors, it was the subsidiary rather than the independent firms which were growing faster in employment terms, and that two of the four fastest-growing independents had been acquired by 1990.

Table 2.9 Age of surveyed establishments on science parks

Age	*Number*	*%*
Less than 4 years	45	28
4–9 years	57	35
10–25 years	43	26
26–50 years	11	7
More than 50 years	7	4
Total	163	100

Source: UKSPA-OU-CURDS survey

There is, moreover, evidence once again of a North–South difference, or perhaps more accurately a difference between a select group of southern and sunbelt science parks and those in both the north and inner cities. It is, maybe ironically, the latter group which seems so far to be host to a higher proportion of new independents. The *Tenants Directory* sample suggests that parks in the north have more new establishments (42 per cent) than those in the south (29 per cent) and fewer relocations (table 2.10). We cannot at this stage know the meaning of these data. They may, for instance, merely indicate their lesser attraction to established firms. But the data do fit with those already cited earlier in the chapter. The possibility is certainly emerging that science parks in different parts of the country may be the site of highly contrasting social and economic dynamics.

Table 2.10 North–South differences in new start-ups (percentages in brackets)

Divide	*New establishments*		*Relocations*		*Total*
North	50	(42)	69	(58)	119
South	28	(29)	67	(71)	95
Total	78		136		214*

* This total is higher than the total in table 2.8 because the data here exist for a larger number of establishments
Source: UKSPA (1987)

Facilitation of R&D links and technology transfer between host academic institutions and park firms

Science parks, by definition, are located on, or close to, academic or other research institutions. At the core of the science-park concept lies the idea that scientific knowledge leads in some linear progression to technological innovation. Universities are seen as repositories of scientific expertise and research, and the view is that the UK is good at basic science but bad at commercialising its fruits. Science parks are a way of orienting academe more closely to the needs of industry.

Science parks are thus in large part based on the premiss that they provide a focus for university–industry linkage, technology transfer, and the application of university research to commercial needs. There are two principal ways in which they are said to assist this interaction:

(i) Academic start-ups – academic staff taking research out of the academic laboratory on to the park, starting up their own firms and moving into the market.
(ii) Tapping-in – new establishments with no previous contact with the host academic institution, and existing establishments relocating units on the park, making use of university resources, expertise, technology, knowledge, and so on.

We shall discuss each of these modes of linkage in turn.

Academic start-ups

If any one firm typifies the model of a university start-up firm located on a science park, Laser-Scan is it. Laser-Scan originated in the Cavendish Laboratories of Cambridge University, where its founders were high-energy physics researchers who developed a scanning device using new laser technology. The instrument was originally constructed in the mid-1960s to assist in the analysis of sub-atomic particle collisions in bubble-chamber experiments. Several machines were subsequently built for other universities, and Laser-Scan was established as an independent company in 1969.

By 1973 Laser-Scan's activity was too great to remain in the Cavendish, and the firm became the first company to move on to the newly established Cambridge Science Park. The firm sought new markets for its instrument outside physics laboratories, and, partly through contacts with the university geography department, developed its technology for application in cartography. Technical assistance was also provided by the CAD Centre at Cambridge.

Laser-Scan has subsequently developed its technology into a range of image digitisers, plotters, display and editors, and gained a Queen's Award for Technological Achievement in 1982. It produced the largest liquid crystal display screens in the world, and its computer-controlled laser-deflection technology had no direct competition anywhere. Recent application areas include displays for aircraft traffic control, and security printing (e.g. banknotes). The fundamental research link with the Cavendish Laboratory has not been maintained over the years, but the present chair, like the founders, previously worked at the Cavendish.

Laser-Scan is thus a paradigmatic academic start-up, fitting well into the classic model. How typical of science-park firms is it?

We have shown earlier that new firms in general are perhaps less common on science parks than might be expected, given the rhetorical emphasis on the creation of something new. But within the total set of start-ups, how many are firms begun by academics? Our 1986 survey found that in fact over half of them were of this type. Of 183 establishments surveyed on this, one in six (17 per cent) of all establishments were university start-ups. The movement of key personnel from academia to become key founders of enterprises on science parks is often emphasised as being a fundamental principle of science-park development. In response to being asked about the most important factors influencing the firm's choice of location on the park, again a similar number (16 per cent) mentioned that a key founder had worked at the local academic institution. Of the 128 independent firms in the sample with available data, thirty-two (25 per cent) mentioned having a key founder at the local academic establishment. These are all impressive proportions and lend support to the model of science parks as sites for the development and commercialisation of scientific break-throughs made within academe.

However, once again it is important to look more closely at the social processes behind the numbers. Both the incidence of academic start-up firms and, more importantly, their social nature vary considerably from park to park. Cambridge Science Park, in spite of being home to Laser-Scan and several other companies founded by academic staff, has a comparatively low percentage of academic start-up enterprises. This, of course, relates to the data previously given on start-ups. In contrast, on Heriot-Watt in 1985, of eighteen establishments, all but three were either university technology transfer institutes, or start-ups with a strong Heriot-Watt academic input. Aston Science Park also has a high percentage of new-start companies, and many of these are academic start-ups. A major factor encouraging new-start firms at Aston is the availability of venture capital from the park's managing company, Birmingham Technology Ltd. This capital fund is used to support firms that locate on the park, on an equity investment and loan basis. It was clear from our interviews with Aston Science Park company directors that the availability of venture capital from BTL had often played a major role in both the establishment of businesses, and the subsequent provision of growth capital (though it should be emphasised again that Aston's commitment to supporting its tenant firms with capital investment is relatively unusual amongst UK parks).

Moreover, another factor influencing academic start-ups in Aston has ironically been the cuts made to the university Occupational Health and Safety Department following the 1981 UGC funding reductions. This department was drastically reduced in size, with academics being redeployed. The department had close links with industry and government organisations, and demand for its expertise continued after its closure. Some of the academics left to form new-start companies (in one case a co-operative), and other redeployed staff set up a part-time business. In an ironic twist to the themes of academic entrepreneurship and university cuts a total of four firms on this science park have been set up to continue the contracts which the university department formerly undertook!

Another form of academic start-up consists of those establishments which remain part of the university, but located on the science park. Heriot-Watt Research Park hosts five Technology Transfer Institutes which are not independent firms, but are administratively linked to university departments, their staff being employed as academic staff. These units are each aimed at providing technical expertise in specific sectors, including computing, petroleum engineering and medical lasers. They may provide consultancy services to industry, or operate as R&D contractors, designing products for clients. They do not, however, fit the pure entrepreneurial mould of the 'start-up' enterprise, since they are not independent of Heriot-Watt university.

Finally, and even less independent, there are on a number of parks, including Heriot-Watt, Bradford, Cambridge and Swansea, establishments which are simply parts of university departments. The university microelectronics laboratories of Swansea and Bradford are located on their respective parks, for instance.

Our purpose in drawing attention to these developments is not to criticise them. Indeed, in many ways they probably hold the potential for much richer and more fruitful long-term development than the classic model (see chapters 6 and 8). But that is the point. They are not the classic 'scientific-whizz-kid-turned-entrepreneur' model. The Aston experience points to the weakness of market mechanisms, at least in some parts of the country, in providing venture capital (we shall return to this issue in chapter 7). It also shows how what looks at first sight like academic entrepreneurship in the commercial sense may actually be the continuation of already existing activities, forced to take on a new guise because of government cuts. Other establishments again are attempts to forge more substantial relations between academic research and other activity on the science park. The point we are making is that none of these conforms to the social process of having a bright idea in the lab and going on to the science park to commercialise it. We shall be questioning later (in chapter 3) the degree to which these results raise question marks against the very model of scientific discovery and technological innovation on which science parks are based.

Tapping-in

We shall use the term 'tapping-in' to mean the accessing of academic resources by science-park establishments. This is a key principle of science parks, whereby firms on the park can build contacts with the host academic institution and benefit from the use of academic resources. In this way small firms which cannot afford to employ full-time research expertise in specific areas, or cannot justify the cost of purchasing expensive scientific equipment, can have access to them within the host academic institution.

To what extent does tapping-in occur?

Perhaps one of the most surprising results of the UKSPA-OU-CURDS survey was the finding that formal research links between academic institutions and establishments on science parks were no more evident than similar links with firms located off-park (table 2.11). Formal research links such as 'employment of academics', 'sponsoring trials or research', 'testing and analysis', 'student project' work and 'graduate employment' were fairly similar for park firms and off-park firms. However, significantly more park firms than off-park firms mentioned 'informal contacts with academics' and the use of academic facilities such as computers, libraries or dining facilities as being important. Indeed, Aston University has set up a special library service for linked firms. The survey further revealed that 60 per cent of science-park firms have used informal academic contacts, 28 per cent of firms used consultancy or part-time academic staff, and 14 per cent sponsored academic research or trials. However, the depth of these links is unclear. Our interviews with park establishments suggest that research links with the host academic institution are usually in firms that already had such links. In many cases even these decrease over time. And firms which move on to science parks with no previous academic research links do not usually forge them. One Cambridge Science Park-based academic said, 'It is easier to "tap out" [to the park from the university] than to "tap in" [from the park] to the university, but overall there is a low level of park–university links.'

Our in-depth interviews on Cambridge Science Park confirm

Table 2.11 Links with the local academic institution (%)

Nature of link	*Science parks*			*Off-park*		
	South	*North*	*All*	*South*	*North*	*All*
Informal contact	67	57	60	42	47	45
Employment of academics	24	30	28	30	26	28
Sponsor research/trials	19	12	14	19	12	15
Access to equipment	35	40	38	21	36	30
Test/analysis in HEI	17	10	12	7	17	13
Student projects	17	25	22	14	31	24
Graduate employment	22	33	30	28	31	30
Training by HEI	2	5	4	5	9	7
Teaching programme	7	5	5	5	14	10
Other formal links	4	–	1	–	–	–
Use of facilities:						
Computer	11	22	19	9	5	7
Library	44	49	48	16	21	19
Recreation	15	36	30	14	2	7
Conferences	11	17	15	7	16	12
Dining	15	25	22	9	3	6
Audio-visual	9	10	10	5	3	4
Other	4	6	5	2	3	3
University as customer	7	18	15	12	19	16
No response	7	19	15	30	22	26
Number of firms	54	129	183	43	58	101

Note: Figures are percentage of firms mentioning each factor as being one of three most important links with the HEI. The final row shows the total number of respondent firms. Firms mentioned more than one factor, so percentages sum to over 100
Source: UKSPA-OU-CURDS

other evidence that the overwhelming majority of firms have no research contact with the university. One classic university spin-off studied had had little research linkage to the university for some years. Dr Bill Bolton, who is both academic and park entrepreneur, states that 'in some respects the direct connection with the university is surprisingly limited and its immense resources have not yet been harnessed to the full by the local high-tech community' (Bolton 1986: 2).

Moreover, the frequency of real research links does not seem to be increasing over time as one might possibly expect with the maturing of the park–HEI relationship. The 1990 follow-up survey of UKSPA-OU-CURDS only contacted, on this question, a very small number of establishments, but the main changes registered between 1986 and 1990 did not indicate depth of research links. Of the biggest increases in contact only student project work (66 per cent) was really to do with research, the other notable increases being in use of recreational (60 per cent), library (73 per cent) and computer (40 per cent) facilities (Storey and Strange 1990). The use of audio-visual and specialist equipment also increased, as did – more hopefully from the point of view of intellectual exchange but probably more a reflection of the dire state of things in academe – the employment of academics part-time (33 per cent). None the less, contacts may increase over time as some parks (for instance Warwick) specify as a criterion for location actual or potential links between the applicant establishment and the local HEI.

Of course this evidence relates to formal links and direct connections. There are many who would claim that informal links are more important, and table 2.11 gives support to this. Not only are these links hard to pin down, however, and in many cases as we have said rather shallow, but also once again they vary very much between parks. Moreover, the 1990 follow-up survey showed no increase in this kind of contact (Storey and Strange 1990). The claim for the significance of informal links is most often made in relation to Cambridge (although Bolton's statement might question their importance even there) but our analysis indicates that the social ambience within which such informal networks can develop is not present to the same degree in all science-park locations. Once again we shall be exploring this issue a bit further, this time in chapter 6.

Employment 'creation'

A further theme emerging from our assessment of their objectives is that parks will stimulate the creation of employment. This was an aim cited more often in the early than in the late 1980s. Job creation has now become an objective mentioned most often by local authorities, which feel they have a clear responsibility to local people. For others it has become a longer-term objective after 'wealth generation' (a term which itself often remains unspecified) through investment in high-technology firms. Employment on science parks has been increasing steadily. As we have already seen (table 2.5), it doubled to 10,540 in the two years to the end of 1988.

There is no doubt, then, that there has been growth in employment on the parks themselves. But this is not, of course, equivalent to employment creation in some sense resulting from the existence of the parks. In other words, these figures in themselves do not enable us to evaluate the degree to which science parks are succeeding as a policy for local employment generation. First, a high growth rate in science-park employment is not the same thing as employment growth in firms established on parks. Total park employment may be growing due to the arrival of new firms. In order to separate employment growth within existing firms from that brought by firms new to the parks it is necessary to look over a period at firms sited on parks. Such data are minimal, but we do have detailed information on job growth in fifteen Aston and twenty Cambridge Science Park firms. This shows, for example, that employment growth in these firms over the year 1985–6 was 19 per cent in both Aston and Cambridge. A further source of information is the follow-up survey to UKSPA-OU-CURDS, conducted in 1990. This showed that in the 63 per cent of establishments still operating on a science park in 1990, employment had risen by a remarkable 71 per cent over the four-year period, from 1,493 to 2,558 (Storey and Strange 1990) (an annual increase of 13 per cent). However, once again it would be inappropriate to derive averages from aggregates, since this growth was concentrated in a very small number of firms. Just ten establishments provided 46 per cent of gross and 96 per cent of net employment growth over the period. None the less, employment growth within existing park firms was significant even though obviously not as spectacular as indicated by figures that include jobs in establishments new to the parks. There is, moreover, a further aspect to this. UKSPA has argued that science parks act to stabilise company location by offering larger premises as firms expand, and during each of the years 1985 and 1986 UKSPA park managers reported that one fifth of tenants moved to larger premises, usually on the park. Moreover, the UKSPA-OU-CURDS survey found a big difference between park firms and the group of similar non-park firms. Expanding non-park firms in their majority anticipated moving site, whereas only 15 per cent of park firms expected to move off-park as they expanded. Second, by no means all establishments setting up on science parks are new. As we have already seen, many are relocations. Such relocations, though adding to total science-park employment, do not create jobs; they simply move them between locations. In spite of the criticisms which have been widely levelled at policies which lead simply to competition between areas, and the patent wider-level costly futility of it all, to a local authority trying to combat a declining employment base and a dearth of jobs in newer sectors, even such relocations may of course be welcome news. Even in this respect, however, the evidence should serve to dampen enthusiasm. As table 2.12 shows, our UKSPA-OU-CURDS survey indicated that, of over a hundred relocating firms questioned in depth on this issue, over two-thirds had come to the science park from within the same city, and a further 17 per cent from within the same county.

Table 2.12 Previous location of relocated park establishments

Previous location	*Number*	*%*
Same city	70	67
Same county	18	17
Elsewhere in UK	15	14
Abroad	2	2

Source: UKSPA-OU-CURDS

If the evidence so far indicates a degree of success, but hedged about with serious reservations, about the degree to which science parks can contribute to local economic growth through direct employment creation, that impression is reinforced when we come to consider the type of jobs on science parks (table 2.13). Compared with the structure of the national labour market, the nature of employment on science parks is unusual. Forty per cent of those employed in 158 park units surveyed in 1986 (UKSPA-OU-CURDS) were qualified scientists and engineers (QSEs). Of the founders of independent firms, 52 per cent had higher degrees. Much employment is thus of a 'professional' nature for people with relatively high academic qualifications. Clearly, these are not huge employment numbers, but neither are they laughably small. Parks were in 1990 the place of employment of perhaps 5,000 graduate scientists and engineers. But this high proportion has its own ambiguities in a labour-market context such as an inner city.

Table 2.13 Employment on science parks, by occupational category (%)

Qualified scientists and engineers	40
Other professional and managerial	17
Clerical, administrative	19
Manual	17
Other	8

Source: UKSPA-OU-CURDS survey

Science parks are also predominantly locations for the employment of men rather than of women. This is true both in absolute terms and on a number of comparisons. Only around a third of the total jobs on science parks are held by women. This is lower than the proportion of female employment in the economy as a whole. But it is also less than the proportion in the control group of similar firms, but not located on science parks, in the UKSPA-OU-CURDS survey. This is significant. There is clearly a gender issue here which needs further investigation.

The level of technology

A crucial aim of science parks is to attract 'high-technology' enterprises that operate at the 'leading edge' of technology. Not surprisingly, then, in a recent survey 31 per cent of science-park establishments considered that they had a 'leading-edge technology' product. A further 43 per cent thought that their products or processes involved the application of advanced technology. One third said their research was 'radically new'. Nine-tenths said they were doing some research and development.

What can we say about the technological level of science parks? Our data suggest that science-park establishments are relatively sophisticated technologically. But rather than being 'leading-edge' in any absolute sense they appear to be more involved in new applications of relatively novel technologies, to be small innovators rather than involved in major innovative break-throughs, and indeed often to operate as diffusers rather than innovators *per se*.

Given the strong popular association between high tech and information technology (IT), it is no surprise that the IT sectors predominate. The 1985 UKSPA 100 per cent survey of park establishments found that 49 per cent were principally involved in IT (computers, electrical and electronics) (see table 2.14). Another 10 per cent were in the biotechnology, pharmaceutical and chemicals sectors. These

Table 2.14 Industrial sectors of science-park establishments (%)

Computers	33
Electrical, electronics, instrumentation, robotics	16
Chemical, pharmaceutical, biotechnology	10
Other industries	6
Consultancy and testing	16
Financial, business services	8
Other, miscellaneous	9
Total	100

Number of establishments: 301
Source: Dalton (1985)

broad-brush data were confirmed in the more detailed 1986 UKSPA-OU-CURDS survey (table 2.15). Over half the 183 establishments surveyed were predominantly in computing and microelectronics. In addition, many others were in sectors closely related to IT, such as instrumentation, automation and technical services.

Table 2.15 Main sectors of science-park establishments

Activities	%
Hardware and systems	26
Software	15
Microelectronics	10
Instrumentation	3
Automation	4
Electrical equipment	1
Medical	3
Pharmaceutical	3
Fine chemicals	1
Biotechnology	3
Environmental	4
Mechanical	1
Design and development	4
Analysis and testing	14
Other technical services	3
Financial and business services	5
Other	1

Source: UKSPA-OU-CURDS

Thus, overall, science parks as a whole do fit in with the popular conception of high-technology enterprises as IT-based. But this is not true of all parks. It is interesting that Cambridge is less focused on IT, having a more diversified range of establishments in other high-tech sectors. Table 2.16 compares the proportion of IT establishments in Cambridge and Aston with those for all science parks. At this stage the explanation of this greater diversity is unclear. It may be a function of the significant presence of Medical Research Council and Agriculture and Food Research Council establishments and linked firms in the Cambridge area.

Table 2.16 Information technology concentration: Cambridge and Aston compared with park mean (%)

Type of IT	*Cambridge*	*Aston*	*All*
Hardware and systems	13.6	27.8	26.2
Software	9.1	22.2	15.3
Microelectronics	13.6	2.8	9.8
Total	36.3	52.8	51.3
Number of establishments in survey	44	36	183

Source: Cambridge and Aston: UKSPA-OU-CURDS, UKSPA *Tenants Directory* and in-depth interviews; all: UKSPA-OU-CURDS

However, sectoral classifications of what comes out of establishments give a relatively superficial idea of the activities going on inside them. For example, the computer hardware sector includes a wide variety of processes, from the manufacture of sub-assemblies, products and systems to the warehousing and distribution of imported hardware products. Table 2.17 shows the principal *activities* of park establishments. Science-park establishments are much less involved in manufacture than off-park establishments but more involved in design and development and software production. The park and off-park units, chosen on the basis of sectoral, ownership and age similarity, were remarkably similar in most other activities. Some differences might be expected. For instance, there are restrictions on manufacture on some science parks. Some of the similarities are interesting, for instance that park units do not seem to be more concentrated on research as a principal activity. Park establishments seem to be more focused on sales activities, similarly involved in servicing and repair and, surprisingly, have more (low-tech) warehousing than off-park units.

The strong warehousing and sales orientation of a large group of park units is consistent with our in-depth empirical research from two parks. Here, some establishments were essentially marketing units for multi-site companies; others marketed bought-in hardware and software with minimal modification by the park firm. One park company was acknowledged in one of our interviews as a 'commercial operation with no innovation ... effectively a sales office'. Another park contained firms that were 'only dealerships'.

Table 2.17 Principal on-site activities of establishments (%)

Activity	Establishment: *Science-park*	Establishment: *Off-park*
Manufacture	14	27
Production/design	6	6
Software	4	1
Design/development	21	14
Research	6	5
Analysis	5	7
Consultancy	5	5
Training	4	2
Servicing/repair	11	11
Marketing/sales	8	6
Warehousing	12	10
Others	5	6

Source: UKSPA-OU-CURDS

Finally, our in-depth interviews suggest that not all establishments on science parks characterise themselves as 'leading-edge'. One firm said it was 'exploiting an American computer software package' that had 'no perceived advantage over competitors'. Another firm was simply 'not a high-technology business'. Software-oriented firms tended to be mainstream producers of software applications, rather than being involved in software R&D. Indeed, one interviewee clearly saw his software firm as a consumer of software developments rather than being part of software R&D.

In examining the relative technological level of individual establishments, one approach is to look at inputs to research and development (R&D). A range of measures are normally used, particularly the proportion of highly qualified scientists and technologists, and expenditure on R&D. There is no doubt that science parks contain many firms engaged in work employing highly educated staff. The percentage of qualified scientists and engineers (QSEs) employed by park tenants is generally very high. Of 157 science-park establishments interviewed in 1986, eighty-eight said that over 40 per cent of their staff were QSEs. In fifty-nine establishments, QSEs represented over 60 per cent of staff employed. In independent firms, over half of company founders had higher degrees. On the other hand, there were twenty-seven establishments with no QSEs at all in their employment.

Another measure of level of technology is financial – usually measured as the percentage of turnover devoted to R&D. Here we found that, of 102 park units providing data in 1986, sixteen spent zero on R&D, thirty-eight spent up to 20 per cent of turnover, twenty spent between 20 and 40 per cent, and twenty-eight recorded very high R&D expenditure, at over 40 per cent of turnover. These latter are extremely high percentages. However, measuring the significance of R&D in small firms is difficult (Wyatt 1985). At the beginning, many firms have few sales. Small firms may have no clear demarcation between R&D and other activities. Some firms specialise in R&D contract work, so high R&D figures could be a manifestation of park establishments as specialists in R&D-oriented activities. Thus parks, as we might expect, could be focuses of an increased division of labour.

Our data thus suggest that science-park establishments have a reasonable intensity of R&D effort. But it is possible to go beyond this a little to look at measures of R&D *output,* particularly patents registered and new products launched. Looking at the patenting performance of science-park units, we find that 28 per cent of our sample had taken out one or more patents in the last two years (table 2.18). There is a significant North–South variation; 41 per cent of southern science-park establishments had lodged patents, compared with 23 per cent of parks in the rest of the UK. Overall, science-park units had a greater tendency to take out patents than did our group of off-park units, 18 per cent of which had patents.

Table 2.18 Patenting activity of park and off-park establishments

	Establishment	
Patenting activity	*Science-park*	*Off-park*
Establishments lodging patents in last two years	51	19
Establishments with no patents	132	89
Percentage with patents	28	18

Although patent statistics are often used as proxy measures of technological performance, there are problems with comparing establishments – the tendency of firms to take out patents varies between sectors, between firms and between countries (Taylor and Silberston 1973; Pavitt 1982). For our purposes we omit those establishments that (i) operate in sectors such as software production and business services, which are not patentable, and (ii) are very new – this to cut out a potentially random element. Omitting these, we find that a high proportion of establishments that can patent do patent – 34 per cent of science-park establishments had taken out one or more patents in the previous year, and 20 per cent in the year before that.

On product launches the data are less clear-cut. There is considerable variation between science-park sites in the tendency for establishments to have launched new products. In some sites over two-thirds of units had launched new products, in others (not necessarily the newest parks) under half had done so, though there is no noticeable North–South difference. Overall, there is no difference between park and off-park establishments. However, it does seem that those establishments lodging patents are more likely also to be involved in launching new products (see table 2.19). Higher levels of R&D expenditure, on the other hand, do not correspond to higher levels of new product launch. One reason may be that some establishments concentrate on research and the early parts of development, rather than the production of new products for the market. Another may be once more that many science-park establishments are relatively new and that products are not ready in significant numbers yet.

Table 2.19 New products and patent activity compared

Patenting	*No new products*	*One or more new products*	*Total*
No patents	96	118	214
One or more patent	16	54	70
Total	112	172	284

Source: UKSPA-OU-CURDS

Overall, the quantitative measures of R&D inputs suggest that science-park establishments have relatively high R&D expenditure and a large proportion of qualified scientists and technologists.

On outputs, the results are less clear. Significant patenting takes place but product launch data are less impressive. However, the commercial launch of a product is rather far downstream in the innovation process and thus not so immediate an indicator. And we have seen that a large proportion of science-park establishments are not engaged in manufacturing products. Again, we get a picture of a varied group of establishments – some focused quite strongly on the early parts of the innovation process (R&D and design), others with a software orientation, some involved in later aspects of innovation – including product launches – some confining themselves to more upstream activities.

Finally, this high degree of variation among science-park establishments was fully confirmed in the in-depth interviews, and the figures give some idea of the range of technological level which exists. Figure 2.2 contains a set of thumbnail sketches of park units. These establishments are relatively sophisticated. There

Figure 2.2 Leading edge? Examples of more sophisticated establishments

- A fast-growing independent microcomputer manufacturer, assembling and marketing a computer hardware product, based originally on a US design, and subsequently on a UK design. The UK design was contracted out to a major electronics firm, but the science-park establishment does some in-house development. This firm has expanded on the park into new, larger premises.
- A biotechnology firm set up to exploit government-funded R&D from government research labs, many of which are located fairly close to the science park. This firm is essentially an R&D management and marketing operation, with most R&D contracted to off-site establishments.
- A company developing sophisticated software which calibrates scientific instruments and other precision equipment. This firm has close relations with the (mostly foreign) suppliers of the hardware, and is regarded as being world-class in its field.
- An R&D division of a multi-site company, which was in the forefront of laser applications a decade ago. The R&D unit was formed following key research done at the university by an employee. The science-park unit continues to do R&D for the group.
- A CADCAM division of a multinational company, with in-house software development. The CADCAM system is based on a system developed in another university (i.e. not the host HEI to the science park). This system has been further developed by the park unit, which also runs a bureau service for CADCAM clients without their own equipment.

are examples of R&D units, of management and marketing establishments, relatively sophisticated software development and systems development. In figure 2.3 we present brief descriptions illustrating the diverse range of less sophisticated establishments. They include sales establishments, consultancies and a company with a large warehousing and sales bias. Overall, while the technological level in quite a number of establishments on science parks is undoubtedly high, the range of levels is perhaps even more striking. What is clear is that, on these measures, only a minority of establishments fulfil the expectations generated by the popular conceptualisation.

Figure 2.3 Leading edge? Examples of less sophisticated establishments

- A consultancy firm advising customers on computing and microelectronics applications, and on exploiting available government grants for the take-up of these technologies. (May move into product later.) No on-site R&D.
- A consultancy which advises clients on the interconnection of microelectronics components. This unit is a division of a larger corporate group, and has no on-site R&D.
- A consultancy in CADCAM, using bought-in hardware but with some in-house software modification.
- A firm based on a software application for civil engineers. Software originated in an HEI (not the park host) but further developed in-house.
- A marketing company in the opto-electronics field.
- A marketing company selling custom chips designed by an associated company elsewhere. Chip testing done in-house.
- A consultancy firm in the pharmaceutical field. No in-house R&D.
- A young firm producing a range of data recorders for industry. Early development done elsewhere but subsequently in-house. Assembly from bought-in components and software development on-site.
- A software applications department of a multi-site firm.
- A consultancy firm in the health and safety field. No on-site R&D.
- A division of a large multi-site electronics company, exploiting MOD-funded R&D done elsewhere. In-house development and prototyping.
- A marketing and sales division of a foreign multinational pharmaceuticals/health care company. Products are researched, developed and manufactured elsewhere.

CONCLUSIONS AND FURTHER QUESTIONS

Our results throw doubt on the relation between the concept of a science park presented in the literature and the causal processes attributed to it. On none of the four criteria investigated do the outcomes simply corroborate the existence, or functioning, of the postulated causal relations.

Now, of course, had the proponents of science parks constructed their concept of a science park and its intrinsic potentiality in a different mode, they could have handled this situation. For much (though by no means all) of the empirical evaluation as reported in the last section was derived from extensive (as opposed to intensive) research. The proponents of science parks could therefore have pointed out that there is no necessary link between causal power and the event which is the outcome of a concrete process. Had they been good realists they would have known that there were now two possibilities to ponder. First, it may be that the causal powers, although correctly diagnosed, are not in operation because of the presence or absence of some contingent factor. It has to be said, however, that while this remains a possibility the evidence so far points in an entirely opposite direction. Rather, we have seen that even in cases where the numerical outcome conforms to expectations it does not confirm in all cases the existence of the hypothesised causes. Once we looked behind the numbers at the social processes which produced them these were seen in some cases not at all to correspond with the popular conception. This was the case, for instance, with some of the new companies generated from within the local university. Secondly, however, there is the possibility that our results may indeed provoke sufficient doubt to call into question the conceptualisation of causal relations and the popular model itself.

In fact, of course, the proponents of science parks are not, in a philosophical sense, realists. The relation between their concept of a science park and the processes which the establishment of such parks are meant to provoke was not of the nature of a causal power which was intrinsic to the phenomenon but which might or might not be realised. What they are postulating is a vague description to which they attach causes which they expect to be visible in effects. Unfortunately for the proponents of science parks, on the whole they are not.

We must, therefore, turn to examine the concept itself. The rest of the book is, in a sense, concerned to do this and on the way to draw out the social and economic implications of the reformulation which we propose.

Exploring these implications pushes us towards a broader understanding of what science parks represent as part of a wider, strategic perspective. One thing which clearly remains a question at the end of this chapter is the validity of the very model of science and technology on which science parks are based. The evidence so far is ambiguous, but certainly throws doubt on the concept of the relationship between research, production and commercialisation as a simple linear sequence. This is an important issue for the whole philosophy of science parks, and we shall take it up in detail in the next chapter.

But models of science and technology are socially constructed and have social implications. Chapters 4 and 5 draw on chapter 3 to examine the social basis of the status enjoyed by these scientists and technologists, and their place within the class structure. One thing which clearly emerges is the contribution of geographical form and spatial symbolism in the construction of social standing. Chapters 6 and 7 build on this and examine the dilemmas for policy which are posed both at local level and at interregional level between the north and south of the country. There are clear differences, which need to be examined and explained, between the sunbelt south-east and the rest of the country. There are contrasts between the aims of the various agencies involved, between public and private sectors, and new kinds of models are being developed. It is clear that the actual social dynamics of science parks are in fact quite varied, and that this variation has a systematic geography. This range of issues is opened up in chapter 6 and taken further in chapter 7, which focuses particularly on the relation between the public and private sectors, and on how that relationship itself varies between different parts of the country.

Chapter 6 takes up the issue of geographical variation at the local level and examines two science parks in very different parts of the country. The information from the present chapter, for instance on employment, raises crucial questions here. If much of the employment on science parks is for male QSEs how does that relate to the local economic strategies of the public-sector authorities which are often, in the inner cities and the north, responsible for science parks and see their main objective as economic regeneration? How do such high-tech islands relate to their surrounding local areas?

The fact that science parks may be 'islands' is an important point. We have seen that science parks are not insignificant as locations for the employment of qualified scientists and engineers. And Lord Young, in his speech quoted earlier, spoke of the problem of the separation of the groves of academe from industry and from wealth creation. But bringing together commercial-sector QSEs and academia implies, in its turn, other separations. It implies, for instance, the geographical separation of the kinds of economic activity on science parks from the rest of the economy; in particular, and quite explicitly stated in the whole philosophy of science parks, it implies the separation of research from direct production. What kinds of further issues, and possibly further problems, does this type of separation raise? Such spatial distancing is a social question too. Are science parks symbolic of the growing polarisation in the United Kingdom today between an élite workforce and 'the rest'; indeed, more than symbolic but actually, through their spatial separation and self-conscious promotion as élite locations, reinforcing of such inequality? That question links right back, of course, to the model of science and technology which in the first place posits a separation between thinkers and doers, between research and direct production. It is to the examination of that model which we now turn.

NOTES

1 Much of the data used in this section are from those published annually by UKSPA. We have used data from 1985 to the present, but principally those for end-1988 and October 1990. The latest data for October 1990 confirm all the trends and conclusions discussed in chapter 2. One science park, East Anglia, has closed. Two others, Reading and Silwood Park, Ascot (Imperial College), are registered as open, with tenants. The North–South divide is similar to 1988, the 'south' with 26 per cent of parks, having 50 per cent of building area, 27 per cent of tenants, 55 per cent of buildings in construction and 49 per cent of employment:

	No. of parks	*Area of building (m²)*	*No. of tenants*	*Buildings: area under construction (m²)*	*Employ-ment*
South	10 (26%)	194,295 (50%)	370 (37%)	28,333 (55%)	7,171 (49%)
North	29 (74%)	192,913 (50%)	642 (63%)	22,995 (45%)	7,537 (51%)
Total	39	387,208	1,012	51,328	14,708

The number of tenant establishments per park has generally increased, but unevenly, with nine parks showing lower tenant numbers:

Cambridge	82	Aberdeen	19
South Bank	77	Salford	18
Aston	69	Birmingham	16
St Johns (Camb.)	62	Sussex	16
Surrey	59	Wrexham	16
Warwick	58	Keele	15
Nottingham	41	Silwood Park (Imperial, London)	15
Heriot-Watt	37	Cardiff	14
Billingham	36	Aberystwyth	14
Sheffield	35	Loughborough	13
Bradford	31	Merseyside	13
Glasgow	24	Hull	12
Durham	23	Reading	12
Swansea	22	Sunderland	11
Manchester	22	Antrim	7
Brunel	21	Kent	6
Stirling	21	St Andrews	6
Southampton	20	Leeds	5
Clwyd	20	Bangor	4
Bolton	20		

The difference between the north and south has decreased slightly as regards average employment per tenant establishment. In the south the average has fallen to 19.4 from 20.0 in 1988. In the north it has risen from 9.8 to 11.7.

2 This is fully borne out by the latest data, obtained as this book was going to press. On Cambridge Science Park the opening of new buildings has allowed an influx of tenants. This includes a new

building for smaller companies. The combination of the relatively small number of parks with the phasing of construction periods makes the usual assessments on the basis of aggregate data particularly difficult.

3 The off-park establishments were somewhat older, less likely to be legally independent but in broadly similar sectors/trades. The science-park establishments were somewhat younger, geographically more concentrated in the less prosperous parts of the UK and, at the time of the 1986 UKSPA-OU-CURDS survey, had lower levels of employment than the off-park firms.

3

SCIENTIFIC AND TECHNOLOGICAL DIVISIONS

THE LINEAR MODEL OF INNOVATION

The model: foundation concept of science parks

At the core of the science-park phenomenon lies a view about how technologies are created. The view is that scientific activities are performed in academic laboratories isolated from other activities. The resulting discoveries and knowledge are potential inputs to technology. Science provides break-throughs from which new technological goods may spring. Thus the science-park model is based on the assumption that technological innovation stems from scientific research. The model is seen as extremely relevant by those who believe that the UK is good at science but bad at applying it towards commercial ends. The argument goes that universities have many brilliant people making new discoveries but that they lack the means or the will to reach out to the market. Science parks constitute a channel by which academic science may be linked to commerce. Thus science parks are there to promote, not 'science', but its application in technology. Science is seen as the basis for a specific kind of technological development.

This, then, is a highly particular model of scientific research and industrial innovation. Fundamentally, it is a linear model, in which there is a chain of successive, interrelated activities. These begin with basic scientific research and pass through applied and more developmental research activities, the development of new product and process ideas, the evolution and testing of prototypes, to commercial production and finally to diffusion.

Figure 3.1 illustrates this linear model of innovation. At its simplest the linearity is from research to development to diffusion of new innovations (top line of figure 3.1). In other words,

Figure 3.1 The linear innovation model

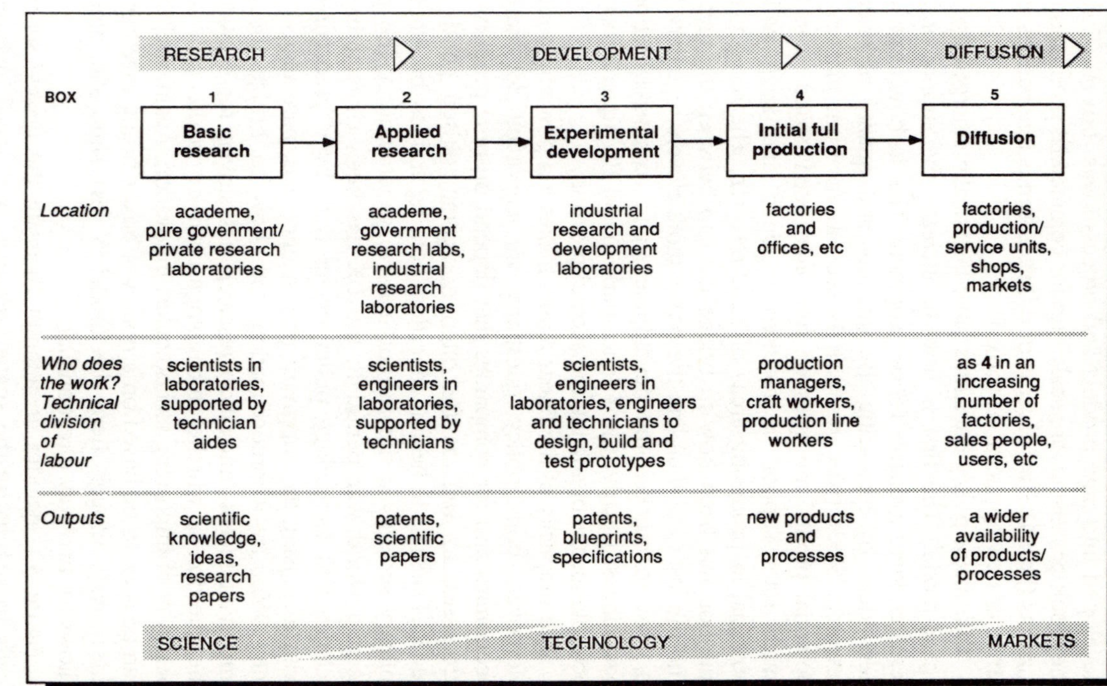

science leads to technology and marketable use (bottom line). More specifically, the model assumes that basic scientific research produces new ideas (box 1 of figure 3.1). Such ideas need to be worked up, using applied research methods (box 2) and developed into new product and process ideas (box 3). These, if they look promising, are scaled up for commercial production and markets (box 4). The process of diffusion sends innovations into a range of new applications (box 5).

Two major policy questions have historically emerged from the linear model. The first is how to increase the supply of basic research ideas available for development. That is, how to generate new ideas by increasing the quantity and quality of scientific research. Since basic research is seen as far from markets and potential profits, the question has often been posed as to how much state support is required to build and keep national potential in promising areas of basic research. The second major question has been how to bridge the gap between the first box and the last (and thus link the boxes more closely together). That is, how to quicken the development and commercialisation of basic ideas. Clearly, the science-park model fits squarely as one possible means of solving the second problem.

The simplicity of the linear model makes it a powerful concept. Moreover, both policy questions are important for national economies and for companies that depend on new products and processes, as well as for often non-commoditised goods and services such as in health, education, housing and last but not least military sectors. At the risk of massive generalisation, the post-second world war period in the UK has been characterised in science and technology policy circles both by a preoccupation with increasing the supply of new science and, particularly since the 1960s, by a preoccupation with how to bridge the gap. Important recent innovation research has striven for answers to the question of whether 'technology push' or 'market pull' is the best way of improving technological innovation. Again, such ideas can be seen to be based on a notion of boxes strung together in a linear way, though here the model is less simplistic, since it allows for two-way directionality.

Figure 3.1 includes an attempt (beneath the boxes) schematically to summarise the spatial, technical and social divisions of labour implied by such a model. For example, basic research is normally located in academic laboratories, government research labs and some basic research labs funded by industrial corporations. These labs fit most closely the popular image of science being conducted by white-coated scientists, supported by groups of technicians. The applied research and consequent experimental development could also be conducted in academic labs, but is more and more likely to be in industrial R&D laboratories. Historically in the UK, as we shall see in chapter 5, the early boxes have been the workplaces of first choice for graduate scientists and for those well qualified graduate engineers who could get in. As we move further along the innovation line, according to the linear model, we move further from the high-status work to work that includes more engineers, designers, technicians, craft and other production workers, and salespeople. Status, historically, has been associated with distance from direct production, sales, and manual work, and closeness to clean, white-collar lab-type environments.

In most linear innovation models direct production does not figure very strongly. Science leads to technology; basic research to development and diffusion. Production is not central. Thus, in the division of labour, production managers, skilled craft workers and production-line workers appear only at the end of the line and after the process of innovation. It is clear that bridging the innovation gap is also a social issue.

In theory, the enterprises located on science parks fit in the second and third boxes – close to academe, bringing ideas from basic research towards commercialisation. They may occasionally fit into box 4 if the first production units are on a small scale and do not involve major, physical production processes – software production is one clear example. They are staffed by high-status groups of scientists and engineers who work in laboratory-like environments.

The science-park model is just one approach to bridging the gap. Obviously, there are others. The most important twentieth-century model for bridging gaps and linking boxes has been the vertically integrated corporation. Corporations like IBM, ICI, Du Pont, Siemens, Bell, Mitsubishi, and so on, have built up basic research, R&D units and all the rest to allow them to compete in producing new electronic, chemical and pharmaceutical, and engineering products for global markets.

One of the difficulties of understanding the full import of this model from within certain cultural perspectives, perhaps particularly the British, is that it seems so obvious. In fact, as we have already said, it is a highly particular view of the process of research and innovation. As we shall elaborate later, there are alternatives. Moreover, even as a description of what happens in the British context, it is increasingly being seen as inaccurate. But, most important at this point in the argument, it is a model which has a number of significant characteristics and necessary social implications. First, as figure 3.1 indicates, the linear sequence of activities implies a parallel *division of labour*; there is a sequence of different groups of people performing the successive tasks. Second, this is an intrinsically *relational* model. That is to say, each link in the chain is defined by and necessarily entails links with the upstream and downstream activities. Each element, for its existence, its purpose, and its very definition, is dependent on the other links in the chain. Third, and putting these two characteristics together, it is a relational division of labour with a distinctly *hierarchical content*. It is the classic model of the separation of conception from execution.

This linear model is the first of three axes in our alternative formulation of the concept 'science park'. The very model of the science–industry relation, therefore, which is an essential part of the conceptualisation of science parks in itself intrinsically entails social inequality. It is fundamental to the arguments we shall be putting forward in chapters 4 and 5 about the class location and social position of scientists, technologists and engineers.

The roots, and cultural specificity

The linear model derives from a particular interpretation of the historical evolution of the increasing role of science in industry. And the way in which scientific knowledge and scientifically and technologically trained staff have been employed has taken a particular form in the UK, with implications for innovation policies – including the science-park model itself.

The relationship between universities and technological innovation historically arises from the growing contribution of basic scientific discoveries to industrial development. Economic histories contrast the early period of the industrial revolution, dominated by artisanal inventors, with the growth of the science-based chemical industry in the second half of the nineteenth century and the electrical industry from early in the twentieth. The example of the discovery of the first artificial dye (aniline) is often used to illustrate some implications for the new type of science-based industry. One such implication was that the ability of the German companies to dominate first the dye industry, then the chemical industry generally, was related to early moves to build strong links between basic research into chemical science in universities and German industrial companies. German universities also grew to produce enough qualified chemists to staff the new factories.

> BASF was in a position to contribute to the development of new chemical processes which required something more than individual brilliance, and depended on sustained cooperation between research scientists and qualified technologists. The establishment of a Department of Chemical Technology at Karlsruhe Technische Hochschule as well as the flow of graduate chemists from German universities very much facilitated their efforts to recruit highly qualified staff capable of product and process innovation. BASF, Hoechst and Bayer were managed by chemists who considered it their business to maintain close touch with the progress of university research ... In 1880, Germany accounted for about one-third of world dyestuffs production, by 1900 about four-fifths.
>
> (Freeman 1982: 30–1)

Put in linear model terms, then, the argument is that German success resulted from pulling close together the early and later stages of innovation. In Germany a division of labour was built which did not completely separate academe from industrial production. Rather, the mental labour component involved strong academic–industrial integration. The combination of high long-term investment in innovation, the development of a wide range of technical skills in industry, and the relatively well integrated division of labour is held up as the reason for that early German national industrial success.

The German example is also often contrasted with the relative 'amateurishness' and 'backwardness' of British industry and universities. It was Perkin, a British chemist at the Royal College of Science (Imperial College), who discovered the dye, but the German companies which grew from its industrial development.

> Perkin's discovery, neglected in Britain, was taken up immediately by the more scientifically-minded directors of the new German chemical industry, and the rapid profits

> accruing from synthetic dyes were ploughed in to create an enormous and dominating German chemical industry. This, though at first ancillary to the textile industry, was, through its capacity for the production of nitric acid for use in the new explosives, to provide the sinews for both the First and Second World Wars.
>
> (Bernal 1969: 631)

> This extraordinary failure [in the UK], despite an early researching lead, has all the signs of the under-investment in innovation, and of lack of qualified manpower to drive a project through to commercial success that we are familiar with a century and a quarter later ... the joint effort of industry and the state secured the professionalisation of science in Germany when Perkin himself was so fearful of the harm done to pure science by his venture into industry that he returned to the academic ivory tower at the tender age of 33. Thus the story of the German aniline dye industry provides a classic contrast between gentlemen and players.
>
> (Rose and Rose 1970: 28–9)

Rose and Rose compare Germany with the United Kingdom and emphasise the historical isolation in the latter of élite scientist 'gentlemen' from commercial and industrial activity – often seen as a more general characteristic isolation of the British upper class from production (Wiener 1985). Engineering was perceived as a low-status occupation associated with dirty factories, and contrasted with the clean, isolated work of the often unsalaried gentlemen scientists.

But the new 'science-based' chemical and electrical industries used graduates, as from the US Institutes of Technology, and German Technische Hochschulen, and increasingly better-educated technician staff. From as early as the 1850s the UK was criticised for falling behind in scientific and engineering education, for not providing paid careers for scientists, and for not developing engineering education, though less was said about the education of technicians and skilled workers.

Already then, by the turn of the century, the divisions in the UK between different types of scientists, engineers, technicians and craft workers were strongly mapped out:

> By 1900 science and industry were distinct social systems, entered by different routes, and with very few institutional

> relationships by which people or information could pass between them ... The gap [between science and industrial production] became itself a new territory, explored, mapped, and eventually controlled by new specialists, the professional technologists, going by the name of applied scientists or industrial scientists.
>
> (Burns and Stalker 1961: 30–1)

Put in linear-model terms, what stands out in the UK is the early separation of academe and industry and thus the highly strung-out nature of the innovation process. The history of science, technology and industrial policy in the UK has continually been a history of how to close that gap.

From 1870, large German companies set up in-house research and development laboratories. United States corporations took on board some aspects of the German industrial research model:

> Germany has long been recognised as pre-eminently the country of organised research. The spirit of research is there imminent [*sic*] throughout the entire social structure.
>
> (Little 1913, quoted in Hounsell and Smith 1988: 3)

In the German innovation process the parts of the linear process were drawn closely together with strong academic–industrial links and at corporate level by putting in-house as many as possible of the boxes of figure 3.1, including a strong R&D unit. This is the element of the German model taken on board most strongly by the US corporations. Leading US corporations recognised the need for scientific teamwork and systematic routine scientific experiments – what Thomas Edison called 'one per cent inspiration and ninety-nine per cent perspiration'. The number of in-house R&D labs grew quite rapidly during both of the rationalisation and concentration waves of the turn of the twentieth century and of the 1920s.

> The development of industrial research within the US firm was heavily affected by the reorganisation of the US corporation during the late nineteenth and twentieth centuries. The growth of industrial research was encouraged by the presence of technically trained managers, a strong central office staff able to focus on strategic, rather than operating, decisions, and the integration within the firm of other functions, such as marketing. The in-house location of research activity allowed for a more efficient combination of

> the heterogeneous inputs necessary for commercially successful innovation. The research facility within the firm also was better situated to utilise and increase the stock of highly firm-specific knowledge gleaned from such sources as marketing or production personnel. In-house research was able to exploit the joint-product nature of manufacturing activity and certain forms of technical knowledge. Industrial research was not only an effect but also a cause, of the development of the modern US manufacturing firm.
>
> (Mowery 1986: 194–5)

As in-house R&D became more and more entrenched, contract research units became less important. In the USA the proportion of total research employment accounted for by independent contract research organisations declined from 15 per cent in 1921 to 7 per cent in 1946 (Mowery 1983).

The approach of the US corporations in putting as much of the innovation process in-house as possible was coupled with strong managerial centralism, the search for giant economies of scale, and the rise of mass-production 'Fordist' production techniques.

> where the manufacturer became the coordinator, his [*sic*] firm grew to great size, and the decisions in his industry concerning current production and distribution and the allocation of resources for future production and distribution became concentrated in the hands of a small number of managers. This centralization of decision making, and with it economic power, was of particular importance because it occurred in industries central to the growth and well-being of the economy.
>
> (Chandler 1977: 372, quoted in Mowery 1986: 212)

Characterisation of the US in terms of the linear model thus needs to include the corporate nature of manufacture. Also, as in Germany, engineers were held in relatively high esteem. But the relationship between academe and industry was more uneven. Some few university institutions, like MIT, had strong direct research links with industry. But higher education was not centrally co-ordinated in the US, a significant proportion was privately controlled, and there was little national co-ordination of policies for scientific and technological education. Still, the production of engineers was much more developed than in the

UK and engineers had a higher status. Some academic departments were used as knowledge and consultancy banks by the bigger companies. But the state-co-ordinated and systematic links of the German model were missing. It was much later, during and after the second world war, that military concern led to the building of strong research links between many of the prestigious universities in the US and the state. Overall in the US the different elements of the linear innovation process were pulled together by the integrated corporation.

The story in the United Kingdom was different. Changes were slower and less generalised. The first world war jolted the government into some action, since the UK was dependent on Germany for specialist lens glass, dyes and tungsten for steel. British Dyestuffs, a forerunner of ICI, was set up by the government in 1914. After the war, the government tried to encourage companies to increase research by setting up Research Associations that were jointly funded by government and firms in a particular sector. In the inter-war years, companies like Ferranti, Metropolitan-Vickers and Lucas followed the large German and US companies by establishing R&D laboratories. Metropolitan-Vickers, a forerunner of GEC, had ten researchers who later became university professors.

The new approach to innovation involved professional R&D departments within firms, and better-organised university research labs. Within firms, qualified scientists were employed both in R&D and in other technical departments. There began to be, at least in a few bigger companies, regular contact with universities (Steward and Wield 1984a). It was not that nothing was happening. However, change was always comparatively slow.

> The British merger wave of the 1920s ... led to the creation of such major modern British firms as ICI and Unilever. The development of the truly giant, centralised firm in Britain thus did not begin until well after the first US merger wave. This slower evolution of British firm structure almost certainly was a significant contributor to the observed differences between US and British industrial research activity during this period. When mergers did occur within British industry, the result in many cases was not a streamlined or efficient structure.
>
> (Mowery 1986: 196–7)

> Chandler's 'managerial revolution' in US business included the absorption into the firm of a range of activities formerly carried out, if at all, via the market. Among these functions was industrial research. In Britain such a transformation and reorganization of major firms occurred later than in the United States, and was less complete. Where research was carried out within the firm, the incomplete rationalization of internal firm structure often hampered its effectiveness. In short, the pattern of development of US industrial enterprises allowed for a more effective exploitation of the complementarities between research activity and production activity. These complementarities are exploited most effectively in a non-market (i.e. intrafirm) setting.
>
> (Mowery 1986: 190)

Given the relative weakness of UK industrial research and its relative separation from direct production, the second world war was a profound experience for those involved in the organisation of science and industrial policy. University scientists were moved into government research, mostly within military projects rather than into the wider industrial economy. The ability of scientists to link up with others in a wide range of goal-oriented projects, from radar to code-breaking using embryonic computers, to the Manhattan atom bomb project, impressed itself on influential scientists and politicians: 'government officials and scientists were equally impressed by the part research and development had played in winning the war [and] saw clearly the potentialities for peace-time development' (King 1974: 13).

These experiences dominated the ideas of those in the USA, the UK and organisations like UNESCO and the OECD which were influential in thinking about the relationship between universities, science and industry, and they helped develop the new area of science and technology policy. The predominant views were that individual inventors had been superseded by large university and corporate laboratories, universities were a crucial basic first stage in a linear process, science now involved big money and that it had to be well organised if nations were to keep up with accelerating international science and technology spending. Just as 'boffins' had been very useful and practical in helping to win the war, so they could remain useful as the first stage in a (linear) process.

> The government attach the greatest importance to science. We recognize the contribution which science has made to the prosecution of the war and the achievement of victory, and we are no less desirous that science shall play its part in the constructive tasks of peace and of economic development.
>
> (Herbert Morrison, Lord President of the Council, quoted in Vig 1986: 15)

These views, also articulated by senior British scientists (see, for example, AScW 1947), still perceived as working at the frontiers of knowledge, were more influential internationally than they were generally accepted in the UK.

Labour governments since the second world war have constantly grappled with ways of closing the innovation gap to encourage stronger industrial growth. There was more state involvement and finance for science and technology in at least four ways. There was increased money for science in universities; more investment was put into research and development in the nationalised industries (like gas, electricity, coal, transport and particularly nuclear energy); more money was spent on civil government labs and on the joint government–industry Research Associations; and there was increased support to private industry, especially to those involved in military work concentrated in aircraft and electrical engineering. However, when the Labour government moved to intervene more strongly, industrial owners reacted:

> So long as nationalization was confined to ailing, inefficient industries or to public utilities, opposition from the major industrial associations remained muted ... But the nationalization of the iron and steel industry was a different matter ... to the FBI [Federation of British Industry] there was a world of difference between nationalization of iron and steel and of an industry such as coal which was sick ...
>
> (Blank 1973: 83, 85)

> The government also sought ... to enlarge the public role in individual industries not scheduled for nationalization ... the attempt to form these Development Councils ran into extraordinary opposition in industry ... Members of the Councils ... would consist of an equal number of trade

> union and employer members ... with independent members and an independent chairman [*sic*] ... The council was empowered to impose a levy for research or the promotion of exports and to undertake a variety of activities to raise industrial efficiency ... The Development Councils proved to be one of the Labour government's least successful ventures ... The major obstacle to the success of the programme was the opposition of industry.
>
> (Blank 1973: 85–7)

Industrial research intensity, measured at four or five times as high in the US as in the UK in the 1930s, was still estimated as three times higher in the US in the 1950s (Mowery 1986: 191).

By the early 1960s fears were also beginning to be expressed about British science. The UK's opting out of the space race, and talk of an academic brain drain to the USA, coincided with gloomy pictures of financial shortages within the universities painted by distinguished scientists like Sir Bernard Lovell, director of the radio telescope at Jodrell Bank, and Sir Neville Mott, head of the Cavendish Laboratory at Cambridge (Rose and Rose 1970: 97). (As we shall see in chapter 6, Mott was later a principal advocate of the Cambridge Science Park.)

These fears were taken up by the Labour Party in opposition, particularly through Harold Wilson's call at the 1963 Party Conference for the modernisation of the economy and challenge to 'forge a new socialist Britain in the white heat of the scientific and technological revolution'. Many of the acts of the Wilson government of 1964–70 regarding science, technology and industry involved attempts to bridge the strung-out nature of academic–industrial links and of innovation more generally. In Cambridge the funding of research institutes like the Cambridge CAD Centre and Research Council units can be seen as the basis of much of the later growth of 'high-technology' industry in the area. Harold Wilson was an early principal political advocate of science parks, writing a letter to university vice-chancellors introducing the idea into the UK.

However, the Labour government which came to power in 1964 was criticised for dividing technology from science by establishing on the one hand a new Department of Education and Science and on the other a Ministry of Technology. The Ministry of Technology also rapidly changed emphasis from technology to industry. As Scientific Adviser to the new Ministry Patrick Blackett explained: 'The problem of the efficiency of British industry, which we were set up to do something about, is much more serious than was generally thought at the time, and a great amount of Government action is needed to put things right' (Blackett, quoted in Steward and Wield 1984b: 199–200).

There was less emphasis on the integration of science and technology into new and old sectors of production than on rationalisation of existing companies through merger. The Labour government's emphasis on corporate inadequacy and in particular on the need for rationalisation and merger had echoes of the US model of corporate vertical integration and Chandler's managerial revolution. There was also, in the 1960s and 1970s, an emphasis on improvement of British management through US-style management education. In the UK, and increasingly in the US in the 1980s, this emphasis has been criticised as further separating top and middle management from knowledge of production processes. Armstrong (1987a) argues that use of the US conception of general management reinforces the 'British disease' of management's distaste for 'the shop floor', which differs from the German and Japanese conception of industrial engineering as high-status work. Armstrong suggests (1987b) that unlike the German view of management as 'management of some substantive process', the typical British idea is of management as a function or status in its own right. Overall, attempts to improve management education have led to practical experience being progressively downgraded.

In the 1960s and 1970s, as in the early post-war Labour government, though state funding was welcomed, any attempts to intervene actively to change the nature of British industry were resisted strongly by industrial owners. So attempts to pull together the stages of innovation in the UK led to further separations. Thus the introduction of management education in universities and polytechnics emphasised general management over management of production processes. And investment in universities, government research labs, nationalised industries, military and nuclear industries, though it produced islands of R&D excellence and career paths for another large group of post-war graduates in science and technology (see chapter 5), always came up against the wall of a significant section of private industry that neither invested nor wished the state to intervene to invest. The separation of science from technology and of universities from manufacturing production continued, even after such strong advocacy of 'white-heat' change. Finally, little emphasis was put on democratisation of, and participation in, decision-making around science, technology and the economy, whether inside government, inside the growing science laboratories, or within industrial companies, private and public.

In the meantime, another phenomenon was emerging in the United States, also strongly based on the need to pull together stages of the innovation process – the research park idea. As we described in chapter 1, the idea grew that close location to research done in universities would help improve the development and commercialisation of basic research break-throughs. And that, rather than requiring direct state intervention, the job of producing innovations was best left to a new breed of sci-tech entrepreneurs. The rapid growth of Stanford Research Park and its environs sprawling into Silicon Valley produced a new variant of the linear model of innovation for the UK and elsewhere – the science-park model, depending more closely on academic research input than the model of in-house research to serve the integrated corporation.

Ironically, in the meantime, the dramatic industrial shake-out of the 1970s and 1980s downswing had produced new ideas about industrial organisation for the future. And these are much less easy to reconcile with the linear innovation model and US models of vertical integration. Rather than the rationalisation, concentration and vertical integration of the large corporation, with the economies of scale of mass production and Fordism, new ideas emerged of a post-Fordist future for capitalist expansion and accumulation. The argument, heavily summarised, is that under 'Fordism' large corporations were able to control a whole gamut of activities, from R&D through production to marketing, to outcompete 'older', craft, smaller-scale production. But increasingly, it is argued (Piore and Sabel 1984) that more specialised (niche) markets are growing with more sophisticated customers (and user companies). Quality and choice are increasingly valued over quantity and uniformity. These changes, together with other economic shifts, including heightened conflicts within the production process, mean that firms must be more aware of market trends and increase the quality and flexibility of their production. Such flexibility depends on specialisation, the argument goes, as individual firms gradually build expertise in filling niche markets.

And this specialisation in turn depends on a new division of labour requiring increased co-ordination between firms. Rather than the large vertically integrated corporate model of constant striving for in-house control, the new division of labour requires more horizontal integration between firms. If they can produce the best-quality products for the smaller niche markets and be able to change production runs more flexibly, then small and medium-sized firms will be able to out-compete the giants. In turn, to keep their markets, large corporations would need to increase their practice of breaking up into smaller business units, each with its own design, marketing, R&D and so on, and with the power to use sub-contractors as well as in-house sourcing. There is some evidence that such changes are happening in some locations and some industrial sectors.

The implied role of science and technology in these new forms of production organisation is quite complex. There is an implied need for more sophisticated products and processes, for better design (both of the look and performance of the goods and of the production processes that produce them) and constant innovation. There is argued to be a need for attention to organisational co-ordination and workforce skills at all levels (not only of management, R&D and design but also of other workers). Japanese and Italian experience is usually cited here and contrasted with the relative decline of US and British manufacturing. Also experience of manufacturing practice becomes crucial for successful management of the production process. Further, rather than a several-stage innovation process beginning with an academic or research idea, which is gradually transformed into finished new goods, the conception is that innovation requires the integration of *all* skills (marketing, design, production and R&D) at *all* stages of the process. The linear model of innovation does not fit such new forms of industrial innovation.

There is already ambiguity in this model. The idea of flexible specialisation and niche marketing does not fit entirely easily with the idea of higher degrees of co-ordination, particularly between different stages of an overall production (including research and development) process. And that ambiguity is heightened by the position of science-park enterprises. On the one hand, in the classic form of the popular conceptualisation they fit in with the flexible specialisation thesis of small, entrepreneurial firms engaged in niche marketing. On the other hand, their very separation, by virtue of their location, from the other stages of the production process, while it highlights a difficulty intrinsic to the flexible specialisation model, also in some ways coincides more easily with Fordist tendencies towards the spatial distancing of the different stages within a highly defined and separated technical division of labour. As we shall see, this ambiguity is at the heart of one set of difficulties which beset the science-park approach.

DIFFICULTIES WITH THE LINEAR MODEL

Science parks and the linear model

The classic science park is clearly consistent with the linear model, based as it is on academe as the source of research ideas to be developed in park enterprises and manufactured elsewhere. Science parks are a 1980s attempt in the UK, using a US approach, to bridge the gap between academe and innovation. They differ from most previous attempts, since they invoke the principle of the individual sci-tech entrepreneur as bridging agent rather than innovation by large vertically integrated industrial corporations.

Yet the evidence so far, presented in chapter 2, indicates that they are not working in the way this model supposes they should. There are two aspects to this. First there is the relatively low level of academic links specifically of the form predicted by the model: academic spin-offs, R&D links, etc. Careful analysis of the situation in US science parks shows that the US model of academic spin-offs has also been exaggerated. The often quoted evidence from Route 128 (Boston) emphasised the importance of academic spin-off firms from MIT (Roberts and Wainer 1968). MIT is certainly one of the locations with significant spin-offs, but even here the study included in its definition of academic spin-offs people who had left MIT up to ten years before starting up their businesses. In a study of 243 'high-technology' firms which started up in the Palo Alto (Silicon Valley) area of California in the 1960s, only eight founders of these firms came direct from Stanford University (Cooper 1971). Moving from spin-off to general academic science-park links of the linear model type, Oakey (1985) has shown that research links between firms and universities have not been as important in California as some of the Silicon Valley histories suggest. Also, a major US National Science Foundation report on university–industry relationships in 1982 found that:

> Of the 39 universities visited in our field survey, 14 universities had owned or associated themselves with industrial parks. Of these parks, only four can be characterised as successful in terms of stimulating technology transfer. However, even in these cases, the presence of the park, in and of itself, did not necessarily strengthen university–industry research programs.
>
> (Peters and Fusfeld 1982: 107)

The lack of such links can be partially explained within the terms of the linear model, although it does begin to point to problems inherent in that approach. Thus the difficulties of building research links are borne out by interview evidence from park management and enterprises. Relocated units with no previous contact with the host higher educational institution find it very difficult to 'tap in' for relevant knowledge and R&D support. Sometimes academic knowledge is thought by park units to be too general for their needs, sometimes too removed from current engineering practice to be relevant. Moreover, empirical results suggesting relatively low park–academic links are perhaps less than surprising when current understanding of how firms maintain and increase their knowledge base is taken into account. Evidence from innovation studies suggests that enterprises need highly specific knowledge in order to solve their problems (Pavitt 1984). Much university research knowledge is either too general, or too fundamental and thus long-term, to be easily usable. This basic research can be conceived of as a bank of knowledge that is a necessary condition of innovation, but by no means a sufficient condition, since highly specific cumulative knowledge is also required. The cost of assimilating knowledge and technologies from outside an enterprise is generally very high. Firms innovate mainly in areas from which they have learned by doing. Thus their knowledge build-up is highly specific and cumulative.

But the second way in which enterprises on science parks are not living up to the expectations of the model is that they are not 'leading-edge'. Rather than producing high-technology leading-edge innovations, chapter 2 showed that park enterprises are better characterised as applying existing 'high' technology to new markets. Many science-park enterprises operate under extremely short-term market pressures, drastically constraining their ability to undertake long-term and risky 'leading-edge' innovation. In practice, many see themselves proudly more as shorter-term and smaller-scale commercialisers and advisers about technology than as leading-edge innovators. But the clear implication, for whatever reason, is that most enterprises on science parks do not in fact occupy that place in the division of labour which the linear model would indicate they should.

These actual characteristics of science-park enterprises raise three important issues.

First, there is the issue of spatial form. While the argument about the gap between basic research in universities and the rather immediate and particular research requirements of individual enterprises was expressed a number of times in interviews, it was also the case that sometimes park enterprises expressed such views about their host university whilst at the same time having on-going research links with academics at institutions away from the science-park area. Such links have often been developed over long periods and are not dependent on very close geographical proximity between firm and academic institution. One firm on Aston Science Park had links with a university over 100 miles away and pointed out that the distance between them would be considered 'chicken-feed' in the US. Another Aston-based firm, involved in computer-aided design and manufacture, does not 'foresee having links with local universities' but does have links with Cambridge, Hull and other universities.

The issue is the relationship between spatial proximity and links of this kind. Spatial form is a crucial aspect of the very definition of science parks, and we shall be addressing it fully in chapter 5. Clearly spatial proximity is not on its own sufficient to generate such contacts. Table 2.11, comparing science-park establishments with those in other locations, was an indication of that. In terms of formal research links there was no major difference between the two groups. There was, however, a significant difference in the degree of informal contact, and this could be important. However, as we shall see in chapter 6, generating the kind of social ambience in which informal contacts flourish and become meaningful is a difficult task, and possibly not achievable in all environments. It may in this context be significant that informal contacts are considerably more important in southern than in northern science parks. Of course, spatial proximity on its own is acknowledged in the UKSPA definition of a science park to be insufficient; a management 'which is actively engaged in the transfer of technology and business skills to the organisations on site' is also one of the criteria for membership (see chapter 2). Even so, as our results show, that transfer is not taking place at any great rate. But that again raises the question of location, and indeed of the very spatial definition of science parks. Given our results, both on academic links and on the technological level of science-park establishments, such establishments require just as strong, if not stronger, links with direct production, design and marketing as with basic research undertaken in local universities – especially since the results of basic university research are more easily and publicly available than more tacit, experiential and commercially patentable information from other sources. There is little scientific or technological reason why most science-park enterprises should choose to locate close to universities, since such proximity is not necessary for their activities.

The second issue concerns science and technology policy more generally. One response to arguments that university knowledge is very general, and that firms require specific solutions to immediate problems, might be to call for universities to reorient their work more closely to the needs of industry. Science parks have been suggested as instruments to promote such a reorientation. But there is concern, even from industrial companies, that the shift further than this along the road to commercial, and thus market-oriented and necessarily shorter-term perspectives, may have negative longer-term effects on British universities' basic research. It has been strongly argued that universities should concentrate on what they do best, basic research, and leave industry to do the more product-market-oriented, shorter-term activities. Thus 'the CBI believes that applied research for industrial purposes is best carried out in industry. We do depend, however, on higher educational institutes to carry out basic and strategic research' (Sir Terence Beckett, Director General of the Confederation of British Industry, 1987: 26-7). Such concern matches that in the United States.

> To try to make universities more like industrial labs will tend to take attention away from their most important functions, which are to be a major source of new public technological knowledge and societies' most effective vehicle for making technological knowledge public.
>
> (Nelson 1989: 240)

The National Science Foundation report, referred to above (Peters and Fusfield 1982), found that universities voiced a number of concerns about university–industry relationships. The most important were issues of: academic freedom, such as the choice of what new fields to explore; research quality and the right to pursue exploratory basic science; and the conflict of interest between public and private funding and between different sources of private funding.

The third and final issue is a social one. Within the linear model the place of science parks is to cover the activities of applied and more developmental research and the development of ideas for new products and processes, and occasionally the development of prototypes. Parks are, in other words, supposed to be quite 'high up' the hierarchical sequence. Yet we have seen that they are not. The companies on science parks are not, on the whole, major innovators on the frontiers of science. This means, in its turn, that the employees on science parks are not in fact as high up in the hierarchical division of labour as the rhetoric would imply. They cannot derive very much status simply from the position which they occupy in the division of labour within production. They are skilled, highly trained white-collar workers; but they are not the scientific whizz-kids just below genius status which the literature would have us believe. Yet our own analysis of the social position of this group in the UK indicates that they actually carry very high status. Chapters 4 and 5 explore why this is so.

Critique of the model

But before we go on to explore these issues, there is a final aspect of the linear model which must be addressed: that this model of scientific research and industrial innovation is not even a particularly productive one. This is the case both in the sense that it is not an accurate portrayal of the way in which the processes take place and in the sense that if strictly adhered to, for instance through policy initiatives such as science parks, it may not be particularly productive of new ideas or innovations.

The linear model was useful in pointing to a relationship between longer-term scientific research and industrial and economic growth, and thus the need to invest in risky research and development (the early boxes of figure 3.1). It was also an important trigger for ideas on how to bridge the gap and make the relationship between

Figure 3.2 Invention, innovation and diffusion as a linear process

Schumpeter has been credited with the key distinction between invention and innovation (Freeman 1982). Whereas invention is 'an idea, a sketch or model for a new or improved device, product, process or system ... an innovation in the economic sense is accomplished only with the first commercial transaction involving the new product, process, system, or device, although the word [innovation] is used also to describe the whole process' (Freeman 1982: 7). Thus there are far more inventions than innovations because only a small percentage of the former complete the innovation process and reach the market. Whilst this definition is invaluable conceptually, we would wish it to include innovation which occurs outside the market place, such as in health. In this broader definition the point of completion of the innovation process would be not market launch but the offering of an available technology to users.

The term 'innovation' is used in two senses. One, as above, to describe first use of a new product, process or system. Two, to describe a process including the activities of research, design and development and the organisation of production of the new product, process or system. The latter is often called the innovation process.

The term 'diffusion' is often used to describe the process of adoption of an innovation from the time when it becomes available to the time of wide use.

Figure 3.3 Product and process innovation

A further distinction can be drawn between technological innovations which are products, such as a novel type of nut and bolt, or aircraft, or a monorail transit system, and ones which are process innovations – that is, innovations in the production system itself, such as the float-glass process or the 'ribbon growth' production of solar cells. The distinction between product and process innovation is useful in drawing attention to those innovations that take place in the process of production rather than as finished products, but it is not always easily maintained. Innovations such as electronic funds transfer (EFT) represent a new service product for bank customers (like through-the-wall cash availability), whilst being driven by process innovation, in this case the electronic transfer of funds and information, driven by the banks' need to reduce the cost of paper transfers.

A further complication is the fact that one firm's product innovation may be its customer's process innovation, as when a machine-tool manufacturer sells a transfer line to a customer.

science and production more efficient. It continues to dominate much science and technology policy-making at international and national levels:

> the traditional model of technological innovation is of a process inaugurated by research, and involving the trauma of development before the climax of innovation. Such a model assumes a convenient linearity which is almost certainly unjustified, but which is also assumed in much government science and technology policy, presumably for the same convenience.
>
> (Macdonald 1983: 28–9)

With the benefit of several more decades of experience and research, it is possible to see problems with early linear innovation theories. Advances on linear innovation theories have included a less optimistic view of the relationship between academe and innovation; a stronger appreciation of the importance of productive activity for innovation; and overall a realisation that the linear innovation model does not nearly reflect the full complexity of the relationship between science and production. There is now considerable evidence, both negatively, questioning the linear model, and positively, suggesting other theories of innovation. Academic scientific research plays an important but more variegated part in these studies. A group of US researchers set out in the 1970s to synthesise the research results on studies of innovation. Their conclusions were as follows:

> When we first began our survey of the innovation literature, it seemed that the process could best be analysed in terms of phases in a linear and unidirectional sequence. ... A simple block diagram would suffice, showing a linear and unidirectional process going from phase one through phase five. The totality constituted the process of innovation. We were disabused of this linear-sequential notion rather quickly.
>
> (Kelly *et al.* 1978, in Roy and Wield 1986: 25–6)

These researchers picked up a whole range of inadequacies in the model. For instance, their review suggested that it is impossible to pin down when exactly an invention happens (that is, the starting point of the whole process) as products and processes become more and more complex.

> As we reviewed the various theories and models, we began to realise that in almost every major innovation of recent times each functional phase is linked in some way to the others: every phase in our block diagram has lines connecting it to and from every other block in the diagram.
>
> (Kelly *et al.* 1978, in Roy and Wield 1986: 27)

Kline (1985), in a major review of the literature on the linear innovation model, comes to much the same conclusion. He believes that the model, simplistic as it is, is still used because no other model has been available. He found that, although there were many critiques of the simplistic nature of the linear model, there were few practical attempts to move beyond it, apart from the two-way linear model of technology push/market pull. Macdonald (1983) suggests that it continues because it is a convenient means of justifying expenditure on academic and basic scientific research. A third explanation emerges from the sociology of science literature (Pinch and Bijker 1989), which suggests that technology has been treated as an autonomous 'thing' that can account for success in innovation. Histories of innovation are focused on the heroic nature of science and on the success that applying science naturally brings – never on the problems and failures, nor with few exceptions on the science that emerges from technology (Vincenti 1984). A fourth reason could be that alternative models are both more complex and require strong empirical backing if they are to replace the simple linear model. However, in the last decade or so, there has been an extremely productive period of knowledge build-up about innovation, allowing for the beginning of the development of new innovation models.

Dosi (1988), in his review of the innovation process, posits five 'stylised facts' on innovation that have emerged from this work. The first three 'facts' follow smoothly from our review of the roots of the integration of science and industry, as follows:

(i) *Innovation involves a fundamental element of uncertainty.* That is, what is searched for cannot be known with any precision, either in a scientific or technical sense, or in a commercial or economic sense. There is uncertainty about which technological directions will be 'backed' with resources (Noble 1984; Pinch and Bijker 1989), as well as about whether 'backed' directions will deliver innovations.

(ii) *Contemporary technological innovation is increasingly reliant on advances in scientific knowledge.* That is, although there is considerable uncertainty, there is usually a range of possible opportunities arising from scientific advances. This allows Pavitt (1984), in his taxonomy of technical change in firms, to characterise one set of firms as science-based, depending on science for their innovative applications. We have described the role of synthetic chemistry in the rise of the chemical industry. Similarly, the electrical and electronics industries depended on electromagnetic and solid-state physics research.

(iii) *The increasing complexity of research and innovative activities militates in favour of formal organisations (firms' R&D labs, government labs, universities) as opposed to individual inventors.* Dosi suggests (as also do Mowery 1983, Teece 1988, and Nelson 1989) that 'the formal research activities in the business sector tend to be integrated within more or less integrated manufacturing firms' (1988: 223).

These three 'facts' could as well fit with the linear model of innovation as with more complex models. The fourth and fifth 'facts', the crucial importance of which has become known in recent years, fit much less well.

(iv) *Significant numbers of innovations are originated through 'learning by doing' and 'learning by using'.* 'People and organisations, primarily firms, can learn how to use/improve/produce things by the very process of doing them, through their "informal" activities of solving production problems, meeting specific customers' requirements, overcoming various sorts of "bottlenecks"' (Dosi, 1988: 223). The implications of the importance of 'learning by doing' for innovations are far-reaching. For instance, there are organisational forms which produce hierarchical labour processes where:

> production workers are treated as thoughtless commodities, supervisors are given the role of non-commissioned officers and the generals orchestrate global production from the top floor of skyscrapers ... Creativity is largely confined to this senior management and an intermediate tier of specialised design workers in their R&D departments ...
>
> (Kaplinsky 1989: 43)

Such forms have been contrasted with organisational forms (in Scandinavia and elsewhere, but most recently Japan) where more integrated labour processes are effective.

There has also been an increased awareness of the importance of tacit over codified knowledge.

> The stock of technical knowledge includes not only what we call the principles of science but also a number of other critical elements. Among these elements are: engineering analyses for specific classes of problems not addressed by science *per se,* codes, practices, know-how, many forms of specialist neuromuscular skills for operating machinery and instruments, and knowledge about controlling and trouble shooting specific processes and systems. To put this differently, there is a vast range of technological knowledge embodied not so much in published literature as in the minds and muscles of many varieties of working technologists.
>
> (Kline 1989: 10)

Vincenti, in his study of production-centred innovation (flush riveting in aircraft) makes the distinction between descriptive and prescriptive knowledge: 'Descriptive knowledge, as the term suggests, describes things as they are. Prescriptive knowledge, by contrast, prescribes how things should be to attain a desired end ...' (Vincenti 1984: 573).

A significant proportion of prescriptive knowledge cannot be comprehensively codified. To this should be added tacit knowledge:

> 'Descriptive' and 'prescriptive' denote varieties of explicit technological knowledge. To these we must add the implicit, wordless, pictureless knowledge essential to engineering judgement and workers' skills. Such tacit knowledge was evident here [in flush riveting innovation], for example, in the ability of workers to form dimples and upset rivets by hand. It appeared also in the attempt by production engineers to forecast the decrease in riveting cost with time and in the 'educated guessing' by which structural engineers arrived at allowable strengths in the face of early incomplete data. Words, diagrams and pictures can help suggest and promote tacit knowledge ... The knowledge itself can come in the end, however, only from individual practice and experience. The fact that tacit knowledge is inexpressible does not mean that it is any the less knowledge.
>
> (Vincenti 1984: 574)

The fact that 'learning by using' leads to innovation has been studied in detail by von Hippel (1988). He and his postgraduate students have studied a range of sectors. He can argue, with

Figure 3.4 The process of technological innovation – an interactive model

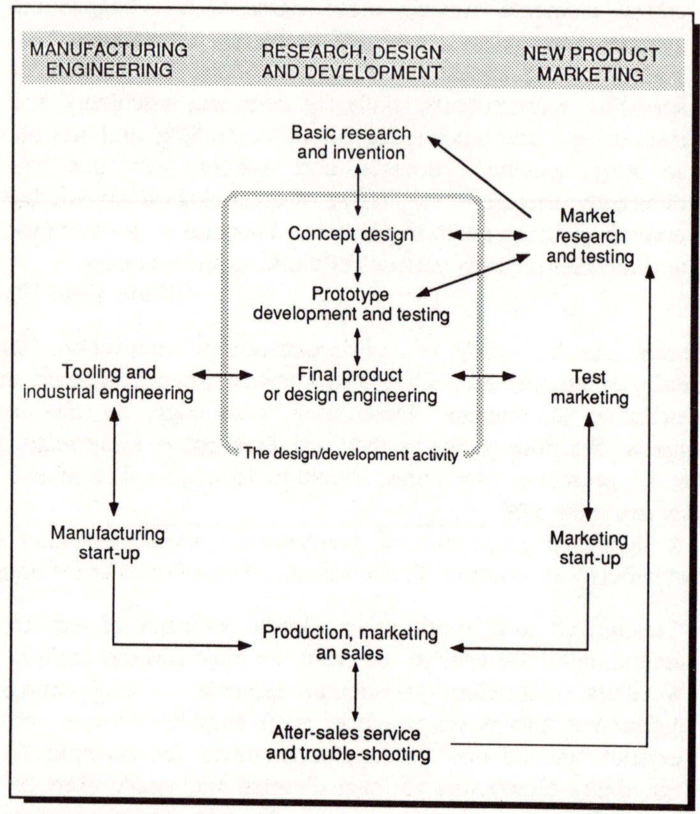

Source: Roy and Bruce (1984)

convincing empirical support, that in a significant number of industrial sectors it is user companies and organisations rather than producer companies that are more innovative.

(v) *Technical change is a cumulative activity.* This fifth 'fact' is a counter to market determinism and concerns more the nature of technological change generally than the innovation process specifically. Technological change is not a simple response to changes in market conditions. Directions of technical change are influenced by the state of technologies already in use. The existing shape of a technology often, or even usually, determines the range of adjustments of products and processes. This 'evolutionary' view of technological change has led to concepts such as technological paradigm and technological trajectory – that there are patterns along which technological change is constrained. That is, future knowledge and practice are constrained by the present (Pavitt 1984). This is so at the level of whole technologies (such as microelectronics and synthetic chemistry) and also at lower levels, so that 'What the firm can hope to do technologically in the future is heavily constrained by what it has been capable of doing in the past' (Dosi 1988: 225).

The most obvious implication of critiques of the linear model is that there can be no one model of the way innovation takes place. One approach, illustrated in figure 3.4, would be to emphasise the role of design, marketing and production, as well as R&D. Kline (1989) emphasises the need for a highly interactive model to capture the intricacy of the innovation processes where ideas can be developed at all points (figure 3.5). It is less easy than with the linear model to describe who does the work at the various points in these models, since it depends on the particular forms of work organisation. For example, the division of labour in the UK would clearly be very different from that in some major Japanese companies.

There are five major differences between linear and more interactive models. First, there is not just one process of innovation, that from research to commercialisation; rather, new ideas are generated and developed at all stages of innovation, including in production. Second, basic research is not the only initiator stage. This is not to imply that basic science pursued in laboratories is irrelevant to innovation. But it is certainly to assert that any suggestion that such research is done in boffin-like isolation is a serious misconception of the processes of scientific and technological work. Third, rather than just being used as the eureka beginning-point of innovation, research results are used, in one form or another, at all stages of innovation. Fourth, the relationship between basic research and commerce is too complex to be understood as a straight-line relationship with a complete divide between phases. In engineering terms, there are feedback loops at all stages. But more, the relationship is also too complex to be understood as a straight-line relationship with a complete divide between mental and manual labour. Fifth, the linear model devalues the contributions of most of the people involved in

Figure 3.5 Chain-linked model of innovation

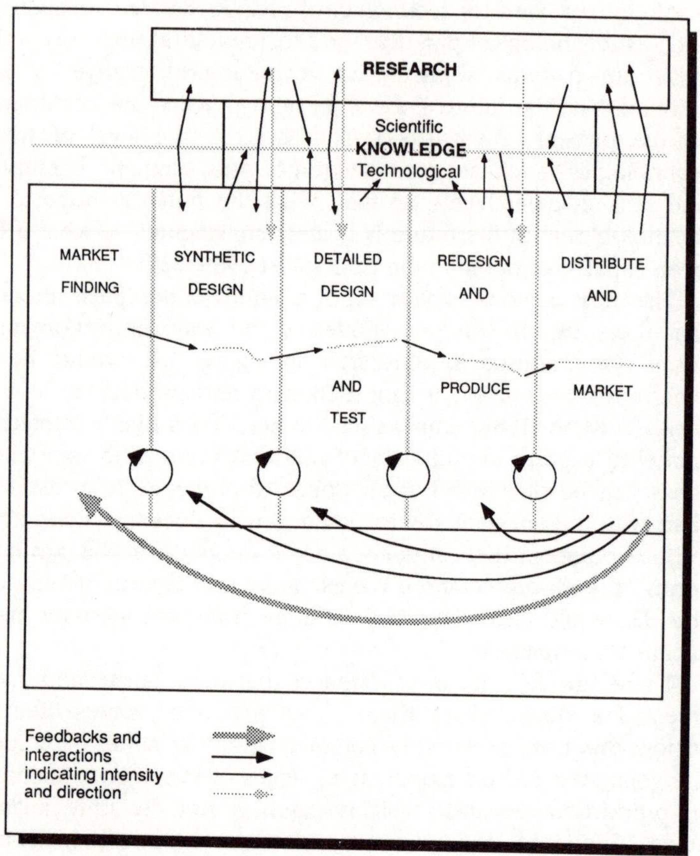

Source: Kline (1989)

innovation, be they professional engineers in design, development and production or those groups involved in setting up and keeping production going. It devalues also the role of technology users, whose ideas and consequent changes of processes and products for their own particular needs can be the starting point of new innovations, small and large.

This is not to say that there is one particular alternative model. Examples of alternatives range from the system-team approach of some innovative Japanese companies to the use of skilled workers in manufacturing innovation in (West) German industry, to experiments in Sweden with more open work environments, to the three-in-one teams of the Chinese cultural revolution. Each approach is different.

All the alternatives, however, share some characteristics which counterpose them to the linear model underlying science parks. They imply far more interaction, and at times real integration, between what the linear model would conceptualise as separate stages of the innovation process. As we shall see, science parks' foundation on such a separation, and indeed the insistence that it is not just a social but also a spatial separation, are potentially problematical. Rather than improving the relationship between academe and 'the real world', the linear model implies a strong division of labour, increasing the separation of mental and manual work. Further, therefore, and of crucial importance to the next stage in our argument, is the fact that none of these alternative views of the innovation process is as hierarchical and élitist in its social implications as is the linear model.

4

ON THE PARK – A VISION OF THE FUTURE OF WORK?

There is a further aspect to the linear model of scientific research and industrial innovation which lends additional characteristics to the imagery surrounding science parks. And as with the linear model in chapter 3, this aspect too has its internal ambiguities.

These characteristics have to do with the organisation and social character of work. In part because of the association, at least locationally, with academe, in part because of lingering notions of curiosity-driven individual 'boffins' as the personnel of the scientific and technological frontiers, the nature of employment on science parks is assumed to be different from the run-of-the-mill job. High-technology enterprises, on science parks and elsewhere, are frequently held to epitomise a style of organisation and of employment which is radically different, in terms of management structure, labour process, and social relations more widely, from that in 'conventional' firms. In the high-tech work environment, hierarchies are supposed to be low and organisations flexible. Work – so the argument goes – is of high status and relatively autonomous and creative. Employees are well qualified academically, and workplaces more like university laboratories than factories. Employees identify their firm's success with their own so that at times they may work very long and inconvenient hours, but they do so with commitment. The resulting enterprises are in consequence successful and can pay well. In such a context, trade unions are unnecessary.

This, then, is a characterisation of the nature of work which is based on the place of these jobs within the overall division of labour, although as we have seen in chapters 2 and 3 this is by no means as high-powered as the supposed place of science parks within the linear model would suggest.

However, there is another side to this imagery. While this notion of a different kind of 'work' is central to the science-park image, science parks in turn are important to the creation of an image of the future of work. Although the number of science-park establishments is small, one major reason why they gain strong media attention is to do with this rhetoric of a style of work which heralds a new future. It is, of course, from the start potentially paradoxical that a set of characteristics which derive from a particular place within a hierarchical division of labour should be held to prefigure the future for the majority, but none the less the issue should be investigated. In this chapter, then, we look at employment, and assess the extent to which science-park enterprises conform to this vision, and also examine to what extent it is – or can be – a prefiguration of the future of work for everyone.

But two other themes are also apparent from the argument in the chapter, and they will run through the rest of the book. The first is that all this imagery is dependent upon counterposition, on establishing its difference from (and utter superiority to) something else. This new, high-technology work culture can be, and often is, counterposed to much of the rest of the economy, especially in manufacturing, and can serve as part of an attack upon it as outmoded. But, second, the 'placing' (literally) of such work on a 'science park' only heightens such imagery and the contrasts on which it depends. The theme of place and the symbolism of environmental context is central to science parks. Their cultural meaning, and the way that is tied to location, is fundamental. It is not for nothing that they are called 'parks'.

The drive to Cambridge

It is, perhaps inevitably, the public presentation of Cambridge and its science park which best exemplifies the message. The dominant imagery is of a new kind of employment set in a classy environmental mix of rurality and ancient (but also very scientifically modern) academe. Media representations of the various routes to Cambridge present different glimpses of it all. Journalists expound in full flow the 'greenfield' character, and the novelty of working in semi-rural areas. The use of the term 'parkland', implying both rurality and the reassuring fact that it is thoroughly under control, is almost *de rigueur*. It is all very far away from the dereliction and drabness ascribed to ordinary industrial surroundings.

> The university city of Cambridge, traditionally a backwater in industrial terms, has become a high technology 'boom town' in a transformation story worthy of a film script.
>
> (Marsh 1985)

> The flat rural landscape around Cambridge, populated by a higher density of small, science-based companies than anywhere else in Britain, seems a long way from the desolate urban areas of the industrial north.
>
> (Marsh 1986a)

> What they want above all is a prestigious environment ... Mr Henry Bennett of Bidwells, project manager of Trinity College's Cambridge Science Park, said this was particularly so when companies were [needing] to attract high-level customers and staff. 'They appreciate pleasant, spacious and restful surroundings and wouldn't go to an industrial type estate at any price.
>
> 'We had a brand new office and warehouse building available on an industrial park only 150 yards across the road from the science park and at half the rent. The high-tech companies wouldn't touch it', he said.
>
> (*Financial Times*, 12 October 1984)

Most visitors to Cambridge Science Park will arrive by car from the south via the M11 motorway. The M11 has shrunk the distance between London and Cambridge – the route from the M25 London Orbital to Cambridge is now a thirty-minute blast, even at the legal speed limit. However, visitors and Cambridge company directors need not set a magnesium-alloy wheel of the Porsche in Cambridge's congested streets when *en route* from London to the science park, for the motorway cuts west of the ancient city to join the fast east–west A45 route which bypasses Cambridge to the north. Slip down off the M11 and head east on the A45 and two or three minutes later the buildings of the science park appear somewhat incongruously. The Napp building glistens white concrete and shining tinted glass. Near by the park's newer brick blocks exhibit classical Greek lintels, round windows and joky details. Exit the A45, drop down on to Milton Road and the science-park entrance is here on the right. London Orbital to science park is thus effectively motorway door to door. Via this route the city of Cambridge, its ancient colleges and its narrow streets appear somehow far away. Indeed, at the wrong time of day it could take twenty minutes to drive from the science park to its progenitor Trinity College in the city centre.

This latter journey traces a different set of social connections, from the university collegiate city centre and the river Cam to the northern edge of town, where the science park lies.

> There are tell-tale signs in Cambridge that things are not what they were. Outside Kings College little has changed: Japanese tourists still congregate in search of the perfect Cambridge snapshot ... Dons in gowns sail past like black swans, students tinkle their bicycle bells ... But on the periphery of the town far stranger sights can be seen. Sleek high-tech offices have sprouted in vacant land; a steel and glass laboratory nestles in lush parkland ... there are signs of an invasion of young, upwardly mobile professionals, followed closely by the aerobics teachers, wine bar managers, boutique owners and tradesmen who cater for their every whim.
>
> (MacPherson 1985)

The imagery of new lab-like work relations is also used in Cambridge Science Park advertisements:

> Three centuries ago Newton was researching in his rooms in Trinity. Many notable scientists have worked in Cambridge since. The fountain of scientific ideas flows here as strongly as ever. Achieving the commercial potential of those ideas – and applying the vast range of local scientific expertise to helping high technology industry – is the aim of the Cambridge Science Park.
>
> (Advertisment for Cambridge Science Park)

and:

> Mr Richard Granger, business director of research and development company Cambridge Consultants, who is trying to recreate the atmosphere of a university physics laboratory in its latest £1m extension to its science park building, puts the advantage of prestigious surroundings in attracting scarce staff rather more forcefully.
>
> 'In the kind of service we provide good staff are fiendishly hard to come by and you've got to play every card you've

> got in the vicious business of attracting them, including the environment', he said.
>
> (*Financial Times,* 12 October 1984)

Another route to the science park is from the Trinity Industrial Estate and Felixstowe docks, like the Cambridge Science Park developed on land owned by Trinity College. Once again, journalists get carried away:

> through the agri-rich counties of Suffolk, Norfolk and Cambridgeshire to Cambridge itself ... Go west from the dock up the near-motorway A45, round Ipswich and Stowmarket and Bury St Edmunds and Newmarket and see the wide fields stretching north through Suffolk into Norfolk, holding beneath their hard ground the arable crops which will spring up in the months ahead ... Out on the M11 just down from the little Archer-like village of Thriplow ... you look for the Porsches. Spot a Porsche, spot a modern Cambridge Man: a hi-tech hot shot wunderkind who talks leverage in the morning (the M11 is so fast), picks up orders on the phone to Boston from his plant in Cambridge Science Park in the afternoon and reviews research into gallium arsenide at the Trinity high table in the evening.
>
> (Lloyd 1986, gender as in the original)

As we have said, this imagery in turn is held to mean that employees are relatively autonomous and creative. They are frequently characterised as having individual commitment to the company, as being highly motivated and competitive. Hierarchies are supposed to be low and status differences weak; salaries are high and trade unions irrelevant.

> People are not on duty just from nine to five – you will find lights burning at all hours of the night. People are not 'managed' in the traditional sense, they are responsible for themselves.
>
> (Professor John Ffowcs-Williams, quoted in Shaw 1984)

It is also, finally and as some of the quotations have already indicated, about lifestyle more generally. Indeed, some of the newer parks might even include a residential component. Meanwhile, an announcement of a new high-technology building near Cambridge city centre proclaims that:

> In an unusual feature, some of the work units will offer not just a laboratory and office space but living accommodation for researchers. 'I understand that some of the people who run these type [*sic*] of companies like to work virtually 24 hours a day – this will give them a chance to live near their place of work.'
>
> (*Financial Times*, 23 August 1984)

From a distance, the science park is simply an isolated group of low-rise buildings. Inside the complex you get the distinctly US feel of horizontal lines of concrete and glass, looking out on to parks and lakes. Some buildings are obviously architects' dreams. As you park your car in spacious car parks and walk into one of the buildings the interior decoration gives a light modern feel ... There is an abundance of potted plants, it is quiet, and in the bigger buildings large reception spaces with comfortable chairs and smiling (female) receptionists. This reception area can be used for 'happy hour' drinks for company employees, as a way of drawing people together and providing some social coherence.

It does indeed feel a long way from some of the manufacturing industry in other parts of the country. And that, of course, is part of the point.

WORKING ON THE PARK

The image is in large part true. Many of our interviews confirmed these characteristics of work on the park. There are indeed some elements of a new organisational style and a new kind of employee that high-technology firms do epitomise.

It is certainly true, for instance, that a high proportion of employees on science parks have academic qualifications. As chapter 2 showed, although there are regional variations, with percentages being lower in the south than in the north, our survey across parks found that around 40 per cent of employees were qualified scientists and engineers. In part, this is due to the nature of the establishments on science parks, their high-technology and research functions, and in particular the comparative absence of manufacturing production. But it is also more than this. Comparison of these figures for qualification with figures for occupation implies clearly that many of these QSEs are in management. In comparison with the rest of British industry, it would seem, management in science-park establishments has a high level of scientific qualifications. In this respect, science-park establishments would seem to be in some measure less vulnerable to the long-established criticism of British industry that there are too few engineers, and poorly qualified ones at that, in managerial positions (see chapter 5). The data on key founders of science-park companies confirm this picture. Of 101 independent science-park companies, 64 per cent of key founders had a first degree, and 52 per cent had a higher degree.

But are science-park companies breaking the mould in wider ways? This section focuses on a range of characteristics often grouped together under the heading 'flexibility'. To what degree, and in what way, do science-park establishments live up to the image of the flexible firms of the future? We also focus primarily on Cambridge Science Park. It is the largest park so far, and the one most often used to illustrate the British science-park phenomenon, held up as a vision of a potential industrial future ... Cambridge's 'science park and the related phenomenon of high technology growth all around the city have presented one vision of what the new industrial Britain could be like' (David 1986). But we shall also set Cambridge against the wider picture of British science parks as a whole.

Our data on flexibility were collected primarily from an Open University-designed 'special questionnaire' which was conducted in lengthy interviews with senior management of science-park establishments concurrently with the UKSPA-OU-CURDS survey referred to earlier. The questionnaire was completed (or at least substantially completed) by managers of eighty-eight science-park establishments, with 1,024 staff, on twenty science parks: eighteen establishments (with 444 staff) were located on Cambridge Science Park; twenty-one (with 245 staff) were on Aston Science Park; and the other forty-nine establishments (with 335 staff) were from eighteen other science parks. In addition, even more in-depth interviews were conducted about flexibility in over twenty Cambridge establishments and a small sample of Aston establishments.

Flexible working time

The key data on time flexibility are summarised in table 4.1. Questions 1 and 2 concern the degree and nature of flexibility in overall working hours, while questions 3, 4 and 5 are about the recording of time spent on individual projects or pieces of work.

Table 4.1 Time flexibility on science parks

1.	Q:	Do some staff work flexible hours?		
	A:	Yes 59%		
		No 41%		
2.	Q:	How do you ensure that staff do a minimum of hours?		
	A:		Enterprises with:	
			Flexible hours	Inflexible hours
		Formal methods (time sheets, fixed hours, target hours, tight supervision)	9%	40%
		Informal methods (trust, high morale, no checks necessary)	91%	60%
3.	Q:	Is the time each employee spends on a project formally recorded?		
	A:	Yes 45%		
		No 55%		
4.	Q:	Do individuals record their own time (or do supervisors)?		
	A:	Individuals 67%		
		Supervisors 33%		
5.	Q:	For which categories is project-time recorded (in enterprises where there is recording)?		
	A:	All staff 58%		
		Some staff 42%		

Source: Special questionnaire; UKSPA-OU-CURDS; further interviews

It is clear from the answers to questions 1 and 2 that, at least in the eyes of these management interviewees, there is considerable flexibility in overall working hours. Fifty-nine per cent of establishments in our survey operated overall time flexibility for at least some of their employees. This is not an overwhelming proportion perhaps, given the imagery, but it is a high percentage compared even with the results of the 1987 Acas survey, which itself found higher degrees of labour flexibility within British industry than had previously been identified (Acas 1988). Moreover, of those establishments which did operate some flexibility in working hours, an overwhelming 91 per cent had no formal controls on the working time of employees. Even in those enterprises where overall hours of working were fixed, 60 per cent used only informal methods to ensure what hours were worked, rather than formal supervision. Phrases used in answer to questions about work control typically included 'We have high enough morale', 'No checks are necessary', 'Trust is enough', and 'Staff must be self-motivating, able to work alone'. All of them indicate that the managers interviewed perceived themselves as giving autonomy and flexibility to their workforce in return for responsibility. Moreover, on all the questions in table 4.1, park establishments indicated greater degrees of time flexibility, and autonomy over it, than the control group of non-park establishments.

There were also some systematic variations, within these overall figures, between different groups of establishments. Thus there was a significant difference between single-plant independent enterprises and non-independents. Independent enterprises (that is, establishments which were not subsidiaries or divisions of larger firms) were more likely to have a more flexible attitude to work time and its observation. Thus, in answer to question 1 ('Do some staff work flexible hours?'), 71 per cent of single-plant independents answered in the affirmative, while the figure for non-independents was only 46 per cent.

This may be related to the other contrast which was readily apparent, this time between establishments on Cambridge Science Park and those on Aston. Contrary perhaps to expectations, it was Aston which lived up to the flexible imagery more. While considerably less than half the Cambridge establishments used flexitime (the precise figure was 44 per cent), on Aston two-thirds of the interviewees reported its use. The Cambridge figure, in other words, fell considerably below the national average for science parks while that for Aston was slightly above it. We have already seen in chapter 2 indications of substantial differences between science parks in the southern sunbelt and those in the rest of the country; and the data here reinforce that emerging picture. The evidence on this aspect of parks will be pulled together and examined in more detail in chapters 6 and 7.

Further into the company the autonomy and creativity continues. In our interviews people talked of coffee on tap, but also of 'Perrier water in the fridge ready for those working on a problem at 2.00 a.m.' (interviews: consultant). 'Key people are often still working at 10, 11, 12 and even 2.00 a.m.' 'At the top level the barrier between work and play has disappeared. People work at what they like to do' (interview: local venture capital). Quite a few employees were dressed informally, though by no means all, some sporting grey or pin-striped suits.

A voluntary, self-motivated commitment to deadlines was taken for granted by the lecturer in charge of the small research company where workers, who 'did it for love, not money', were not paid overtime, but all 'tended to work for enthusiasm' – if the project was exciting, they would work late into the evening. Similarly, in another company, although hours were standardised at nine to five, 'employees were told that they might be required to work more than this ...'. They did get 'peaks of activity', usually in the summer. The firm 'expects people to be motivated enough to do this and to see it as part of their job'. And further light is thrown on the apparent lack of flexitime on Cambridge by the fact that a number of establishments which in formal terms had fixed-time systems pointed out that staff often worked well beyond their formal hours. 'When work needs to be done, people do it', said one MD. And a personal assistant reported, 'The hours are nine to six officially but I often do more. I often go home at six to feed the cat and then return later at nine p.m. to work in the evening', and 'We don't want to lose the friendly community atmosphere. We work hard to preserve the newness. Some people hark back to the good old days (two years ago) when we just had seven people'.

In one firm, a small subsidiary of a fairly large corporate group, the business manager told us he had been at work since four o'clock that morning. A lengthy meeting had just finished at 2.00 p.m. when we arrived, and he looked desperately tired. He and his co-director, who had been at work since 6.00 a.m., accepted long working days as the norm. When asked why they didn't recruit more staff to ease the workload, they thought the idea preposterous: the subsidiary was built around these two individuals, who were indispensable and could not be substituted.

Many science-park units have some employees with this version of flexibility – the flexibility of working on an interesting problem and the freedom to work on it twenty-four hours a day during the peaks, to go off to play squash in the middle of the afternoon, etc.

When it comes to time flexibility within the work process (rather than overall hours) the same picture is evident of relatively high flexibility but with some reservations (questions 3, 4 and 5). In 55 per cent of establishments time was not recorded on individual projects. None the less, on 45 per cent there were systems of formally recording time spent on individual pieces of work. Simply recording the time an employee spends on a project does not, however, necessarily imply lack of autonomy. Any person or organisation might record time spent on a job in order to do accurate costings for clients. The follow-up questions on who records the time spent are more interesting, since they may indicate some level of supervision.

More often than not, respondents said that recording the amount of time worked was the individual's own responsibility. But in a surprising number of establishments (33 per cent of those where project time was recorded) it was superiors who were responsible. Such superiors included: director, production foreman, production manager, departmental head, PA to general manager, technical manager, and chargehand. A significant proportion of staff were thus subject to monitoring from above. However, once again there were variations. Establishments owned by other, larger firms had a higher level of monitoring of their employees than did independent firms. Moreover, and here we begin to pull out an important sub-theme, there was a significant division within the categories of staff who were subject to monitoring. Two distinct categories of staff were subject to the most time recording: firstly there were higher-status groups directly involved with work for clients that was often accounted for on a day-pay basis, staff such as consultants and development and software engineers; secondly there were staff like maintenance engineers and shop-floor production operators. We shall return below to the significance of these divisions.

Moreover, even where time recording is done only for accounting purposes, the fact of such accounting, especially in the context of contract work, imposes its own disciplines. Thus one consultant employee noted that although his work was not monitored in any direct sense, every consultant knew that each day of their time was sold to a client and that if they went off at a tangent for too many days clients would eventually want to have some report on why. He mentioned this as the main form of control over his work, which was 'otherwise free from interference from managers'.

Organisational structure, status and functional flexibility

This relatively high degree of employee control over work time, and the autonomy and responsibility it was held to reflect, were mirrored also, in many establishments, in loose organisational structures, relatively flat status hierarchies and high degrees of functional flexibility.

Almost all the company representatives interviewed, as well as the property management officer of one of the science parks, stressed the 'open', 'organic' nature of their organisation. Once again, self-definition took place frequently through counterposition to something it was assumed would be generally disliked in the new era which science parks represent. Park establishments were contrasted with the traditional and much larger monolithic organisations in which production and research were combined, and which were accordingly rejected: 'we don't want to be another ICI' (a counterposition which was unfortunate, since ironically ICI is one of the major British companies where senior management at least in some divisions has self-consciously and explicitly grappled with organisational stagnation and experimented with the introduction of open and organic structures – Pettigrew 1985). The open organisational structure endorsed by the science-park company representatives can be boiled down to two major elements – the attempt to abandon hierarchy and introduce an interdisciplinary or 'matrix' approach to the organisation of work and a single-status policy both on terms and conditions of employment and in relation to the working environment. Thus one manager stressed both the company's 'very organic management strategy' and also the attempts which had been made, not always successfully, to extend identical policies on job autonomy, performance assessment, etc., throughout the firm.

Many establishments, including the bigger ones, had taken on elements of single status. Even the biggest firms had just one canteen. Only nine of seventy-eight establishments had separate pay structures for salaried and hourly paid staff. Fully 88 per cent had unified pay structures. Further, almost half of all establishments reported that there were no highly defined job descriptions for any category of worker within their organisation (see table 4.2). Again, this is relatively high in comparison with surveys of flexibility in industry more generally. It is also higher than in the non-park establishments for which comparable data were available.

Similarly, there is a relatively high level of organisational involvement of employees in decision-making with regard to, for instance, managerial, product–market and general company policy. In fifty-nine (87 per cent) of sixty-eight establishments where management reflected on this matter in detail, it was argued that staff were directly involved in at least some levels of decision-making. The degree and nature of involvement varied from 'The company is a co-operative; there are equal says and equal rights', 'There is great involvement; we are very open', 'Meetings and consultations involve all staff', to 'Staff are encouraged to discuss all matters informally, except for financial affairs', 'Staff are fairly well involved', and 'All are involved in smaller decisions; management take the big decisions'. In very few cases did the managers interviewed think there was little involvement. Most of these were subsidiary companies: 'Employees have little involvement; policies are determined by head office'. Yet even in the subsidiaries of multinational corporations (MNCs) there was some evidence of more flexible organisational forms. One specifically noted that its science-park subsidiary was 'probably unique' in the company in having single status, no overtime and no hourly paid staff.

Table 4.2. Functional flexibility

		Establishment			
		Park		*Non-park*	
		No.	*%*	*No.*	*%*
Q:	Which staff have highly defined roles?				
A:	All, or a majority	19	26	10	31
	Some categories only, of which:	23	32	13	41
	Secretarial	(10)		(1)	
	Production	(2)		(9)	
	Sales	(4)		(4)	
	Programmers	(2)		(4)	
	R&D	(5)		(1)	
	Managers	(5)		(4)	
	None (highly flexible)	31	42	9	28

Source: Special questionnaire

This relatively high level of reported involvement is spelled out by the data in table 4.3. None the less, some reservations must be registered before over-hasty conclusions are drawn. First, these are the views of management, who might be expected to have a more benign view of organisational structures. Second, the same data also show that in only 41 per cent of establishments was this involvement guaranteed through formal structures of staff–management communication; in 59 per cent of cases it relied entirely on informal methods. While such 'systems' may work well, or even better, under some conditions, they are clearly more subject to both abuse and disuse. These issues will be taken up again in the section on trade unions later in this chapter.

Table 4.3 Participation in decision-making

		No.	*%*
Q:	Are staff involved in policy-decisions?		
A:	Yes	59	87
	of which:		
	Relatively formally	24	41
	Informally	35	59
	Yes	59	87
	of which:		
	All staff	40	59
	Some staff only	19	28
	No	9	13
	Total	68	

Source: Special questionnaire

As regards organisational structure, it was said by one manager that a conscious attempt had been made 'to avoid mechanistic organisational forms'. Thus alongside the traditional line management there existed a matrix structure of interdisciplinary project management, which for the personnel manager represented the essential philosophy of the company. Despite the existence of a degree of graded hierarchy, with group managers reporting to a senior management group, there were within the general scientific/technical area no gradings or other formal divisions. Rather there existed a large 'pool' of engineers and technicians not specifically allocated to any identified work area.

An example of a company in which an organic management structure was almost a definitional feature of the organisation was one which consisted entirely of one small team of eight workers, a mixture of scientists and technicians, working on highly specialised projects. This was headed by an academic from the university, personally committed to making the kinds of connections between academic research and industry that he argued were 'unusual for university lecturers'. The team's concentration on only two or three intensive and innovative research projects at a time made it, however, a paradigm of the organisational forms upheld by other companies on the park, in that the structure was almost necessarily open and interactive. Thus the workforce in the laboratory was described as 'self-motivated'; the leader sometimes had to 'raise excitement', but people were mainly very loyal and found it very exciting to work there. Thus there was a strong link between labour process, organisational structure and attitudes to work.

The deliberate inculcation of interdisciplinary organisational forms was less evident in another of the companies, where the production director, 'as yet unconvinced that any other functions more efficiently', had forced the reintroduction of a traditional managerial structure against overall company policy. However, there was in fact a high degree of autonomy for workers even within this structure, with teams in sales and marketing having to make their own decisions. They were trained to understand company policy and to take initiatives, although they did not formulate most company policy. The exceptions were where a particular individual had specialised knowledge of a specific technique and this could be very important to decision-making.

The clearest expression of the logic of 'open' organisational structures came from the property management officer of one of the parks, although this was far from being the major concern. Significantly, the success of the park was seen entirely in terms of property – the value of the land had been 'colossally increased'. The draw of the science park was essentially a matter of 'prestige', and this was seen as inherently linked with its policy of a move away from direct physical production towards specialised research – criteria for entry into the park included the association of companies with R&D and their advancement of projects which required 'a heavy load of brainpower'. Large-scale repetitive production was actively discouraged – 'they can do that anywhere'; the concept of the park was based on the movement of scientific and technical ideas. Criticising a traditional 'management culture' in which managers were not educated in science and technology, the officer set out the conception of the alternative organisational structures found in the park. These were presented in terms of how 'the US rather than the UK model of management' prevailed in the park, with 'single status, high salaries, "perks", etc.' the norm. 'Informal interaction, informal dress, a non-hierarchical structure' were all features of this type of organisation.

The many similar organisational features mentioned by the managers can be brought together in conveying the ethos of these companies as a kind of 'corporate individualism' – an interactive, 'democratic' approach in which spontaneous commitment and co-operation at all levels of the organisation are seen as appropriate to the kind of small-scale, variable, innovative work on which many of the companies are engaged. At the same time, those involved are assumed to be entitled to 'concomitant' – i.e. high – rewards. This was evident in the recruitment policy of one company where scientific workers were seen explicitly as 'human capital' – 'What we're selling is people'. Thus '... we have to treat them well ...' but at the same time what the company was buying in taking on new employees was expected to yield a high return in terms of commitment and responsibility: 'If you treat people in that kind of way they'll respond – if you're careful in your recruiting. We wouldn't want people who aren't flexible or who are just "takers", taking as much as they can for as little as possible'.

Flexible pay?

Characteristics of responsibility and control, discipline or incorporation, could also of course be encouraged in the workforce by the form of the pay structure. One aspect to note at this point is the quite considerable presence of people working freelance. In our detailed investigation of employment on Aston and Cambridge we found an astonishing 23 per cent of all people with an employment relation with enterprises on the park being employed on a freelance contract basis (calculated as a proportion of the total workers where all, including part-time, full-time and freelance, count equally as one employee). The proportion was similar on the two parks (21 per cent on Aston and 25 per cent on Cambridge). These people are professionals, and the category includes consultants associated only very irregularly with particular firms and others with very regular work. Such a result has its ambiguities. On the one hand it reinforces an important aspect of the 'flexible' character of these establishments. On the other hand, such forms of organisation do impose their own external discipline on workers, removing it from the realm of trust and personal commitment, and they also militate against the notion of the establishment as an organic unity held together by personal bonds and common purpose: such a mode must be far more difficult to maintain if employees are constantly dropping in and out and also have contracts with other companies.

Table 4.4 The payment of overtime (percentage of establishments)

	All science-park establishments	*Science-park independents*	*Non-park establishments*
Paying overtime	25	15	56
Not paying overtime	75	85	44

Source: Special questionnaire

There was, on the other hand, a relatively low tendency to pay overtime amongst the establishments we surveyed. The payment of overtime would suggest that staff require a formal incentive to work extra hours, and would go against the frequently repeated view that overtime is worked voluntarily, indeed willingly, by those in science-park establishments ('working into the night on difficult problems', and so on). And indeed three-quarters of the companies surveyed did not pay overtime, with the proportion being even higher (85 per cent) for independent companies (see table 4.4). Moreover, as table 4.4 shows, these percentages were very considerably higher than in the control sample of otherwise similar firms located off a science park. To the extent that such figures reflect the existence of commitment and self-motivation, science parks again live up to their rhetoric.

Again, however, as in the case of timekeeping, there were differences within the staff. Those establishments which did pay overtime did so to two distinct types of employee: those involved in a direct servicing role, such as maintenance and sales engineers, who are called out on an irregular basis, and – much more predominantly – those in 'lower grades'. In our interviews, these latter were variously described as 'our skilled manual staff', 'production operators', 'shop-floor staff', 'lower grades' and 'those below professional level'.

Only one science-park establishment mentioned a profit-sharing scheme as part of moves towards flexible payment systems, a figure which is very surprisingly low, given the image of entrepreneurial spirit associated with science-park high-tech firms. However, one of the larger firms on Cambridge Science Park had two forms of 'incentive' which could be used to enhance salaries. The first was a profit-sharing scheme in which 'the emphasis is on profit, not on shares'. A proportion of profits was taken at the end of the year, from which 50 per cent was allocated on a *per capita* basis and the rest distributed according to levels of performance within the payments system – thus there is some connection between effort, or merit, and reward. This bonus was paid at Christmas. There was also a productivity bonus scheme, paid quarterly, which was based on the measurement of company performance in terms of time sold to clients. Above a certain threshold of hours sold a bonus was payable, so that people could see the direct influence of their effort – if they put in more hours at a certain point they would be eligible for the bonus. Once again, the issue of pay is strongly related to time and functional flexibility.

The levels of flexibility and commitment expected of employees were also essentially linked with the companies' policy of 'selling themselves' on meeting time constraints set by their client organisations. They would thus 'do whatever they could to get jobs done', 'unhampered' as they were by demarcation disputes, negotiation over details of work organisation or other traditional 'obstacles'. This resulted, of course, in the absence of collectively negotiated agreements over pay, hours or other terms and conditions of employment.

Such nonchalance over employee compensation for overtime was a product of the companies' essentially individualistic pay structures, in which any form of collective negotiating appeared to be 'a thing of the past'. In two Cambridge Science Park enterprises the salary limits for the workforce as a whole were decided unilaterally by a central management group, while individual salaries within these limits were based on personal assessment by the employees' immediate supervisors, backed up by large doses of Management by Objectives (a self-assessment scheme). In neither company did management see any potential difficulties with this approach. There was an appeal system in one company but it had produced no major upheavals; people had complained, but problems were invariably settled – 'sometimes with more salary, sometimes with more explanation'. Again, the level of involvement of employees was held to explain their positive attitude towards the system – everything was said to be 'very open'.

The lack of collective representation of employees was, of course, justified in terms of the level of reward and consultation the companies were capable of offering, and this was used as the basis of the argument against trade unionism.

Trade unions

In the imagery of new work, trade unions have no place. As expected, unionisation was at a very low level. Only six establishments out of eighty-eight had some union members among staff (table 4.5), and of these six two were in the public sector and one was a co-operative. Two establishments had sufficiently clear (if naïvely unacknowledged) opinions on the relevance of trade unions to science-park firms to reply that the issue of whether they had any trade unions or trade unionists was 'not applicable'! On one park the lack of unionisation was a matter of 'quiet pride'. In this manager's view there was a general feeling that 'we don't need unions', because employees are 'well treated' and feel they are in 'a good environment'.

Table 4.5 Trade-union membership

Enterprises with	%
No trade union members	90
All employees members	2
Some employees members	5
Didn't know	1
Not applicable	2

Source: Special questionnaire

The same argument in terms of the company taking over, as it were, the role of the trade union in catering for workers' interests was expressed by one of the managers, who provided some of the most explicit comments on the approach of such companies to trade unionism. No one in his company, he argued, had 'felt they had a need to be in a trade union'; if a ballot of employees showed they wanted a trade union he would be disappointed, because 'it would show we had been doing our job wrong'. He had asked about the issue in the past and overall opinion had been against, but then 'they were quite well looked after'. Going further, however, the manager expressed the view that 'trade unions would be a most damning thing in this type of organisation' – it would be counterproductive. When a company was non-unionised average salary increases tended to be better because people worked hard as individuals, wanting to get out and do their job. The union benefited only those who didn't want to work. Without a trade union people who wanted to work would do well, but when confronted by a 'trade union block' management would take a harder position.

The relationship between the individualism of research expertise and the flexibility crucial to companies which saw themselves as innovative emerged even more clearly in the comments by the same manager on his employees' likely response to the concept of collective organisation. Trade unionism, 'living in the past' as it still was, would be 'difficult' in a situation where most employees were of independent mind. They were all scientifically based people, wanting to do their own thing. He concluded that in the Cambridge Science Park in general there must be few companies which were unionised, because these were professional people. You were dealing not with people doing a routine job, but with people with responsibility, of independent mind – 'I stand up by my own performance'.

The same view was presented by another of the managers, who felt that 'With the openness of the organisation we don't need it'. Again there was the citing of the company's ability to offer 'sophisticated' consultation networks connected with its organic management strategy, along with generous financial rewards, as an argument for the 'unnecessary' and implicitly old-fashioned status of trade unions. The absence of trade unions was linked to the company's 'degree of flexibility' and the fact that it would 'move much quicker' than traditional companies, owing, as mentioned earlier, to its lack of demarcation-style hold-ups. This was also, it appeared, sometimes an attractive feature of the company to customers, including the government.

In the discussion of trade unionism, then, a link was again apparent between the high personal rewards to which qualified workers were seen as entitled and the other facet of this individualism, which emphasised flexibility, commitment and responsibility. Both these aspects were integral to the nature of the scientific workers' labour process.

The flexibility debate

There has in recent years been a major debate, in academic literature, policy studies and the media, about whether or not there is a major structural shift under way to more flexible working practices (see, for example, Atkinson 1984; IDS 1986; LRD 1986). The definition of flexibility has been, indeed, flexible and that has contributed to the high degree of confusion in the debate. However, some characterisations of 'the flexible firm' are now systematically referred to in the debate in the UK. The Institute of Manpower Studies distinguished between numerical flexibility (casualisation, part-time and temporary working, etc.), to which is related flexibility in working times (changes in overtime and shift patterns), and functional flexibility (breaking down existing demarcations between job categories, extending ranges of skills, etc.) and distancing (the externalisation, or buying-in, of products and services). Other classifications have been similar, focusing on the difference between numerical and functional flexibility, or the LRD (1986) distinguishing between flexibility of task, time, numbers and earnings, and NEDO (1986) between numerical, functional, time and pay flexibility, and harmonisation between manual and non-manual workers.

Much of the debate has been about how flexibility has been used to break down existing forms of organisation which have been turned by workers into a source of strength. The aims of 'flexibility' are to increase productivity, to break down demarcations, to 'tap the gold in workers' brains', and so on. As Pollert (1988) has argued, 'capital's harnessing of the inherent flexibility of human labour is the defining characteristic of labour power. It is, and always has been, essential to capital accumulation. How this flexibility was organised has been, and is, part of class conflict' (p. 70). In the early days it was capital which tried to impose rigidities – the opposite of flexibility. What we see today is a re-emergence of a rhetoric of flexibility – though quite how widespread the actual phenomenon really is is very difficult to detect.

The import of the survey results on science parks is twofold. First, this is flexibility, not just of core workers (as opposed to peripheral, in the language of the debate) but of an élite within the core. We found very little in the way of numerical flexibility in the form of casualisation and short-term contracts, although there was the considerable presence of freelance professionals. It was functional and time flexibility which were most apparent. Functional flexibility in particular has been argued to be an emerging characteristic of the 'core worker', securely employed and multi-skilled, though the general evidence for its occurrence is ambiguous (IDS 1986). The functional flexibility which exists on science parks is linked to a high degree of autonomy within the labour process, and with the 'reward' of high levels of participation, at least on an informal basis. The time flexibility was not imposed by superiors, on the whole, but was at least in part a reflection of commitment to finishing a job, interest in the work, and so forth. Indeed, it quite explicitly had its reflection in comments on recruiting: 'careful in your recruiting – we don't want people who aren't flexible or who are just "takers", taking as much as they can for as little as possible'.

The elements which we have looked at reinforce each other and provide conditions for each other. The openness of the organisational structure draws on the high degree of autonomy and individualism within the labour process. The same characteristics of the labour process can lead to a very individualistic labour market in which people are sought after for highly specific skills, characteristics and knowledge. That in turn reinforces the arguments against collectively negotiated wages and conditions, and so forth. And, together, this bundle of characteristics can be held out as a powerful 'new' image of work.

The second import of the characteristics of work on science parks is that the power and significance of this image as a political weapon are further reinforced by its explicit contrast with the 'old-fashioned' image to which it is counterposed. Co-operation, intrinsic involvement, autonomy, commitment to research goals and so on are not to be rejected – they are positive aspects of this work environment. The way in which they are sometimes used, however, is to generate an ideology of work in which individualism is seen as supreme (see also chapter 7) and in which privileged research workers are kept in clean-cut, careful isolation from the contamination of 'nineteenth-century' influences such as trade unionism, or even physical production. Collective solidarity is set up as dull, 'sunset' manufacturing as dirty and old-fashioned, the customs and practices of the labour movement as a clamp on creative individualism, demarcation rules as preventing people being flexible, using their full potential, getting the job done.

PROBLEMS WITH THE FUTURE: ON THE PARK AND BEYOND

Our brief survey showed, then, that life on the park can be good. There are high rewards, high status, a degree of autonomy and control over your own labour process, work in which you are interested.

Internal ambiguities

Yet even here there are ambiguities. First, it is often difficult to assess how real some of the autonomy and flexibility really is. The lack of systems such as profit-sharing, the fragile informality of the forms of participation, provoke further questioning. Second, even without formal disciplines or structures of managerial control, the work can be highly pressurised. The timing of projects (by superiors, or not), the fact that deadlines and keeping to them are an important mode of competition between firms, the high degree of freelancing, all impose a discipline of their own without the need for formal controls. At its crudest, such a work ethic is associated with a desperately competitive individualism. There is pressure to conform to the image. There were occasional references to those who 'didn't fit in'. 'I am amazed at the number of young men in BMWs who leave the park at 5.30 in their cars. If you don't do 90 m.p.h. on the ring road you are nobody' (interview: senior scientist and entrepreneur) – though quite what they were doing leaving at 5.30 is unclear! Indeed, these behavioural characteristics were built into the workforce. More generally, many references were made to 'not selecting people who didn't want to work', but by this was implicitly meant *extra* hours at work.

There is considerable anecdotal evidence that people overcompensate for flexible work conditions. The managing director of one of the larger establishments, with seventy-five staff, explained, 'We work with flexible hours but all our staff work more than the minimum'. Another commented that 'nine to five are official hours but people do much more'. Again 'the lifetime of high-tech firms is short – the pace of change is fast, firms go bust if you don't change with it'. 'Staff are told they have to be flexible. The kind of people who work here are responsible ... they tend to overcompensate.' 'You need a cool personality under pressure.

But once we got an ex-Cavendish Lab guy – he was too laid-back, kept stopping for tea breaks.' None of this is exclusive to science parks, of course. As one interviewee commented, 'Everywhere in [MNC] people work late; there is nothing special about this park', and most academics in universities and polytechnics these days have to work late nights and weekends if they want to get any research done. But there was evidence that the commitment which it is so necessary to demonstrate (and demonstrating commitment can sometimes be as important as the real content of the work done in those extra hours), and which can be associated with much longer and more intensive working conditions, and perhaps in particular the competitive pressure, can lead to overcompensation and burn-out. The time flexibility, which at first sight seems to be so frequently on these élite workers' own terms, is in fact also imposed through competition between individuals, and through pressure to complete contracts.

As Pollert (1988) says, the '"new" types of working offered by new technology and non-standard forms of work offer possibilities of flexible working time, and the much sought blurring between work and leisure' (p. 66). But it is a blurring which, even here, is dictated by the demands of the (paid) 'work' end of the polarity. 'At the top level the barrier between work and play has disappeared. People work at what they like to do. This means lots more bachelors and broken marriages. More single men will stay single' (interview: local venture capital).

Paid work can take over from the rest of life.

Exclusivity

And this in turn means that this kind of 'flexibility' is not generalisable. Indeed, its exclusivity is visible even on the park. For one thing, the managers and QSEs tend to be, with monotonous regularity, young(ish) white males. One park manager thought that 'the ideal person is a bachelor, living at home with a brother and sister. They may have to fly to the US for months at a moment's notice. One person came in for his first day and was sent to Denver for five months. It is common to be away for four and half months'. In some jobs bungalows are rented when contracts are in other parts of the country. 'Three boys [engineers] are sent down, with a woman sent in to cook and wash'. The flexibility of hours is flexibility to suit the demands of paid work;

that commitment which means staying late struggling with some fascinating technological break-through has implications for the lives of others. Much of the excited phraseology which is used in praise of this organisation of work quite clearly places paid work at the centre of these employees' lives. It certainly goes against arguments for shorter working days and shorter working weeks. In all our interviews only one person (one of the few women) mentioned flexibility in positive relation to domestic life – and then it was a case of 'nipping home to feed the cat'.

There are two separate issues of exclusivity here. In the first place, the jobs are designed in such a way that whoever does them will need servicing by others. By definition, then, those people cannot be part of the 'excitement'. In the second place, given the current sexual division of labour in domestic work, this design of paid employment determines that it is in many cases women who will be excluded.

In our survey, out of 138 key founders only five (3.8 per cent) were women, and the few women who had reached top positions had become legends. In our detailed survey of Aston and Cambridge only 10 per cent of QSEs, and 11 per cent of 'other professional and managerial staff', were women (see table 4.6). In contrast women made up a predictable 87 per cent of clerical staff.

Table 4.6 Women working on Aston and Cambridge

Category	*% Female*
QSEs	10
Other professional and managerial	11
Skilled manual	0
Semi-skilled manual	10
Unskilled	0
Clerical/secretarial/administrative	87
Other	6

Source: Special questionnaire

And even in enterprises where status differences (almost always associated with gender differences) were reputedly weak, not all was as it seemed at first sight. In one company with eight employees it was explained that there were no secretaries, and that all technologists did their own secretarial and administrative tasks. But when we arrived to do our interview we were received by the only woman employee, whose desk was situated, conveniently, nearest the door. It became apparent that not only did this highly qualified engineer act as the firm's receptionist, she also had responsibility for inputting the firm's accounts on a spread-sheet package.

There are other exclusions, too, which operate in a different way. There are some even among those who work on the park who are not included in the dominant version of flexibility and the perks which go along with it. As was documented in chapter 2, and further indicated by some of the data in this chapter, even on science parks not all employees are QSEs or managers (table 2.13). And those who are not are often excluded from the kind of flexibility discussed in the last section. Thus some large firms on Cambridge Science Park employ semi-skilled workers as packers. The time of these workers is not flexible, and they are paid overtime. Other firms employ skilled production-line workers who work at a bench all day on repetitive tasks. Managerial interviewees frequently, and sometimes with complete insouciance, revealed the cleavages which exist between different types of worker. 'Open management is more conducive to higher-level work.' 'Staff here are all potential collaborators – except secretarial staff.' 'All staff in firms on the park are treated as part of the campus community – there are no assembly workers.' One manager stressed the company's attempts to extend job autonomy throughout the firm: 'Support staff are expected to run their own functions and to be good at it; this is sometimes difficult with part-time staff such as cleaners but the strategy is still seen as worth while'. And of cleaners in one company where only they did not 'fit in', and while in general employees were covered for six months' sick leave every year, this had to be withdrawn from the cleaners, who 'abused the privilege'.

But neither was it a simple opposition between 'flexibility' for some but not for others. Rather it was that the meaning of flexibility varied between groups of workers. A report on the training needs, for ordinary workers, of new-technology industries in the Cambridge area captures the ambiguity of the term. On the one hand it stresses the importance, among other things, of 'flexible and adaptable working skills'. On the other hand it stresses that prospective employees should appreciate the need for low rates of absenteeism. No time off here for squash in the afternoon. Absenteeism is also flexibility, of course, but it reflects the needs of the worker rather than the employer. Although the data are by no means conclusive, table 4.2 showed that it was secretaries who were most often mentioned as having tightly defined roles. Finally, it was among women and thus among lower-status employees that part-time work was concentrated (see table 4.7).

Table 4.7 Women and part-time work

Women as proportion of	%
All full-time employees	26
All part-time employees	52

Source: UKSPA-OU-CURDS

The conditions of work enjoyed by élite workers on science parks are, therefore, not generalised. Nor are they generalisable. Most simply, as has already been seen, the flexibility in the lives of these employees imposes constraints on the lives of others – those who service them. But, further, the character of enterprises on science parks cannot be generalised throughout the country. Some science-park enterprises are small, and are mainly engaged in particular types of activity with a high 'craft' work content – examples are software production, consultancy, laboratory design and development work, pilot and small-scale assembly work. There is no deterministic relationship between technology, labour process and organisational form, but there are clearly bundles which go more easily together. In this, science-park enterprises are not significantly different from similar units off-park. But it is also that some jobs have intrinsic interest and the possibility of high pay-offs (of a variety of sorts) and of career progression. For people with this kind of job the extra commitment is not only to the company; it is also to the benefit of their own careers. But only some kinds of jobs offer these possibilities. In science-park enterprises those employees whose jobs do not offer these possibilities, like skilled and semi-skilled packers and assemblers, or clerical workers, or cleaners, are in a minority, and can thus be characterised as marginal in comparison with the graduate employees, and the significant proportion of highly qualified 'professional' technicians and administrative employees. This is a very different situation from that which exists in bigger plants and in the economy as a whole, where skilled manual, semi-skilled and unskilled workers still make up a large proportion of employees. Extending the kinds of flexibility enjoyed by a high proportion of science-park employees to this much larger group of workers would certainly not necessarily be a negative thing for them. It would, however, have its ambiguities. Most obviously, extension of the ideas of individual competitivity and anti-collectivism in these circumstances would weaken employees' bargaining position still further than it has already been weakened in the years since 1979. Indeed, the ideology of 'new work' according to science-park myth and practice is: no trade unions at all, not even 'single unions'.

Moreover, it is clear that on this model of flexibility there is a close relation between autonomy, non-unionism and high wages. Interviewees often emphasised that pay was higher in their enterprises in recompense for the harder work of individuals. The absence of trade unions was linked to the degree of flexibility in the companies and the fact that they could move more quickly and individually to reward people for hard work. This element of 'cash bribery' for individual 'flexibility' and 'commitment' was articulated strongly. But, for the majority of workers, the principle of rewarding employees for work flexibility with significantly higher pay has not been a significant element of industrial relations in the period of recent Conservative governments. On the contrary, the 'new' ideology of work relations more generally has been that British workers are 'over'-priced, and that they must accept lower pay rises. It is clear that buying work flexibility from employees on a large scale in the UK goes against one of the major tenets of this economic ideology. The 'new' work conditions which science parks are held to typify are not generalisable in the context of such national policies.

More than that, indeed, the existence of the science-park imagery (and reality) is used against other workers who are counterposed as backward in their collectivist unions and defence of 'old' working methods; and the privileged flexibilities of the QSE are used to legitimise casualisation and insecurity elsewhere. Chapter 3 showed how the linear model was both relational and hierarchical. Here we see that both the reality and the imagery of work on science parks are so too. Science parks in these senses are inherently dependent on counterposition and contrast with 'others'.

Above all, these contrasts are founded on a particular division of labour within work as a whole. Chapter 3 demonstrated that the work done on science parks does not fully live up to its science-park image. But science parks are still élite places within an overall division of labour which is based on hierarchical separation. Certain occupations require and convey skills and status characteristics which are, within the labour market, positional goods (Elson 1987). Their value depends on how generalised they are. In this case, and we shall see this even more strongly in chapter 5, their value is directly related to their exclusivity. 'In a market economy means of production and labour power are positional goods in the sense that the satisfaction yielded by them can be diminished by congestion i.e. by too many others having the same equipment, or raw material, or skills' (Elson 1987: 15). And, in a passage which links this argument with Pollert's point that the battle over 'flexibility', the defining characteristic of variable capital, is a long historical one, Elson characterises thus the current state of play:

> technical progress depends on people trying to change parameters, not accepting that there is no alternative; high levels of productivity depend on people exercising imagination, initiative, and forethought, even on the most routine assembly line; people need to feel some attachment to their occupations, some pride and satisfaction in their job, if standards of quality are to be high. The class answer to this is to allow a privileged part of the workforce – entrepreneurs, managers, scientific researchers, university lecturers, etc – to exercise the initiative and imagination; and to require the mass of the workforce to be passive adaptors.
>
> (p. 18)

The conditions of work on science parks – with all their internal ambiguities – are specifically not generalisable. The status attached to them precisely depends on that being so. And, as we shall explore further, their spatial definition reinforces their social separation. Exclusivity is, after all, what science parks in their archetypal model are all about.

5

SCIENCE PARKS AND SOCIAL STRUCTURE

Whatever the reality behind the rhetoric on science parks, the new breed of scientists, technicians and engineers can certainly lay claim to high status within UK society. There are times indeed when they are given almost mythical significance as the class of the future. William Rees-Mogg (1987) calls them the new electronic class: 'The new class is the electronic class, whose work is individualistic, often incentive-rewarded, personal rather than impersonal, and related to communications which span the world'. Michael Meacher suggests that 'it is the technocratic class – the semiconductor "chip" designers, the computer operators, the industrial research scientists, the high-tech engineers – who hold the key to Britain's future' (1987).

In this chapter we explore some of the things which lie behind the construction of this status. Chapters 2 and 3 pointed to the way in which status can be related, through the division of labour, to the very model of science and innovation which is socially dominant. But they also pointed to the fact that most establishments on science parks are not so close to the élite end of the linear model as is sometimes claimed. None the less, as chapter 4 verified, these are jobs in which conditions of work are considerably better than the average and to which is attached a significant level of social standing. In this chapter we take this exploration further and examine the class location of scientists and technologists and the changes which have taken place since the end of the 1970s. We also analyse the wider social context, and changes in that too, which have added further to the mystique around the practitioners of the new high technology.

THE CLASS LOCATION OF SCIENTISTS AND TECHNOLOGISTS

Some history

The class location of scientists and technologists has been the focus of long debate and also, perhaps, of particular difficulties within a Marxist framework broadly defined. Taken together, the literature emphasises the multidimensionality of the class location of a group, internally diverse, which cannot easily be placed within either of the two main classes in a capitalist society.

Analysis of the growth of white-collar work in the 1960s and 1970s broadly produced two theses. The first, associated with Mallet (1975), Braverman (1974) and Gorz (1967), was that increasing routinisation and de-skilling of white-collar work, including technical work, was leading to its full proletarianisation. For Braverman, the de-skilling of craft work was the strongest form of degradation; and in the UK Cooley (1972, 1980), a practising engineer, has taken the ideas of Braverman into the British craft tradition. He defends the long practical and theoretical training for craft engineers, and attacks the tendency in engineering to divorce mental and manual labour, subordinating manual to mental. This latter concern links in also with the second thesis, associated with Poulantzas (1975) and Carchedi (1977), which insisted on a class divide between the 'new middle class' and the working class. Poulantzas argued that technicians, if undertaking a role of co-ordination and supervision in relation to other workers, were acting on behalf of capital and could not therefore legitimately be included within the working class.

Wright's work has similarly been concerned with the class location of the middle layers, and the evolution of his theoretical stance and the debate surrounding his approach have particularly revolved around the position of skilled workers such as scientists and technologists. In his first formulation (1976) he placed such employees in a number of different locations on the 'class map' which he developed. Thus scientists who are self-employed or are themselves employers would fall on the dimension between bourgeoisie and petty bourgeoisie (let us call this dimension B), those in management positions fall along dimension A, between bourgeoisie and proletariat, and those scientists, engineers and high-level technicians whose employment allows them a high degree of individual job autonomy fall on dimension C, between proletariat and petty bourgeoisie. Each of the dimensions expresses gradations along a distinct criterion: A is concerned with the management and supervision of other workers, B with ownership and C with job autonomy.

Figure 5.1 Wright's (1976) class framework

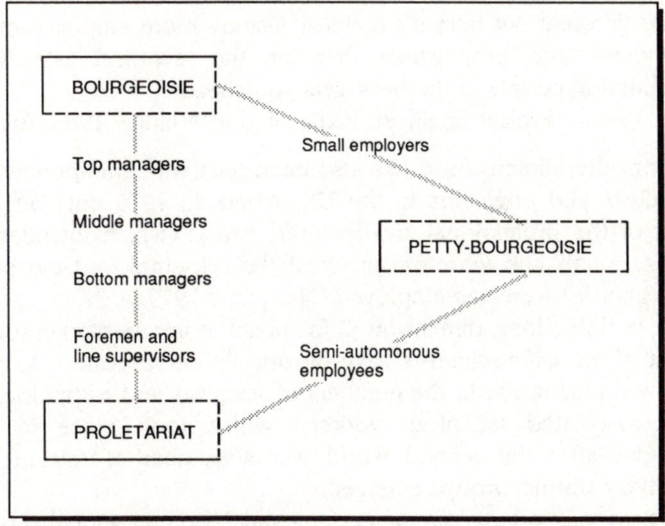

Source: adapted from Wright (1976: chart 8)

This early framework is useful for a preliminary characterisation of the historical placing of scientists and technologists within the British social structure. There are three features which distinguish the British situation. First, there has historically been a comparatively low presence of scientists and technologists on dimension A. In some continental European countries, like France and (West) Germany, engineers routinely engage in the supervision of manual labour. In the UK and the US they do not (Whalley 1986). In France, for instance, advanced design and technical management may often be combined in a way which is extremely rare in the UK (Whalley and Crawford 1984). In the UK the hierarchy in traditional industry was built up not in terms of technical expertise but by movement *away* from technical work and into management. Engineers who aspired to higher career positions had to take up managerial work, a prospect which was by no means universally relished:

> Ideally, I'd still like to be an engineer but paid accordingly. There shouldn't be a big difference between management and the technical side. Chief engineers are involved technically as well as doing management. One wants some management. I'd like some control, but I don't want to avoid all engineering activities. Chief engineer is the highest I could or would want to get. At the other companies it may be different but here it's political. There's more emphasis on memos and propaganda than on the technical aspects. Stabbing people in the back gets you ahead.
>
> (Senior Project Engineer, R&D, cited in Whalley 1986: 105)

Historically, dimension B has also been relatively unimportant for scientists and engineers in the UK, where in 1975 only 5.5 per cent of the professional and scientific sector (which includes all professionals and therefore many of the self-employed old petty bourgeoisie) were self-employed (Marquand 1979).

It is thus along dimension C (semi-autonomous workers) that most of the UK's scientists have historically been located. As part of the dramatic rise in the numbers of scientists and technologists, and associated technical workers, which took place in the decades after the second world war a number of related but relatively distinct groups emerged.

There was a long, strong pecking order by educational qualification. It went roughly from graduate mathematicians and physicists, who were supposed to be able to move into any kind of scientific and technological work, through chemists and biologists, to engineers; from electrical engineers, through mechanical and chemical to production engineers. Then there were those with more practically based HND-type qualifications, recognised as of pass-degree (rather than honours) level. Continuing down the long list, we come to other qualified technicians, to design technicians, and last to craft-trained workers. One interesting characteristic of this hierarchy, and which again distinguishes the British situation from that of some of the continental countries, is the craft apprenticeship system, which for long was a significant force acting against any major structural divide between the new technical occupations and existing manual ones. There were status fragmentations, but there was no clear divide between manual skills and abstract knowledge (Smith 1987). Even more important, the expansion in the numbers of technical workers involved the movement of craft workers to draught and supervisory functions. These are points which will be taken up again later.

None the less, there was a clear occupational stratification of graduate scientists and technologists. Moreover it formed a hierarchy which was strongly related both to degrees of work autonomy and to the perceived distance from direct physical production. In this it mirrored, and related to, the linear model discussed in chapter 3, and its evolution in the UK, and the 1976 class map of Wright. If the craft apprenticeship system helped bridge the gap between manual and mental, the hierarchy none the less remained between them. The hierarchy of jobs awaiting those emerging on to the labour market in, say, the 1960s ran as follows. First, at the top, a large proportion of graduate scientists and engineers were working as academics in higher education, a sector which was expanding rapidly in this period. Second, large numbers worked as research and development scientists and engineers in public research labs, like the United Kingdom Atomic Energy Authority (UKAEA), the National Physical Laboratory and the National Engineering Lab. Third, others worked in private and nationalised corporate R&D labs, such as the GEC Hirst Research Laboratories in Wembley, London, the various ICI labs, the steel, coal, gas and electricity corporate research labs, and many more. Some worked in civil service-type conditions in Research Associations sponsored by the government and companies, like the British Iron and Steel Research Association (BISRA), the leather research association, and PERA, the Production Engineering Research Association. In these three broad areas, technicians worked in support of graduates, though this supervisory function on the part of the latter group was not of a kind, nor of sufficient importance, to push them on to the general managerial dimension A. The fourth broad area of graduate employment was as design engineers in industrial companies, and the fifth, even closer to production, was as production engineers.

In these last two areas of employment, graduates were nothing like so differentiated from other technical occupations. The closer to industrial production, the less clear was the separation between differently qualified groups of technological workers. Certainly, as Whalley (1986) makes clear, the work of production engineers in the UK could not be equated with that of manual workers, and to a considerable degree the differentiation rested on the responsibility and autonomy granted to the former:

> Engineers have the freedom of the factory ... They are responsible for the design of valuable machinery, for preparing estimates that are critical to the companies' survival, or simply for negotiating with manual workers about the daily practices of production. Some of these jobs require more technical skills than others, but all of them are important to the company and someone has to be trusted to carry them out without the close control that shapes the life of the typical manual worker.
>
> (Whalley 1986: 59)

On the other hand, production engineers' conditions of work were often much worse than those of R&D laboratory engineers.

> Production managers objected strongly, as did development engineers, to the proposal that they share a new building. Their present establishments were miles apart, the development engineers in an elegant, pleasantly situated building, surrounded by lawns and gardens and attractively decorated and furnished. The production shops and managers' offices were, by contrast, crowded and makeshift. 'It's often been suggested', said one production manager, 'that we should make up a party of our people and take them over to see around the labs – good idea to let them see what the firm is doing in that line, and so on. Well, that party hasn't come off and won't. We daren't. What the eye doesn't see, the heart doesn't grieve over'.
>
> (Burns and Stalker 1961: 188)

Nevertheless, attempts by professional engineering institutions to demarcate graduate professional grades from non-graduate technical engineers and technicians often seemed a long way from the reality in production, where those with the highest qualifications did not automatically have more autonomy, or authority over those with lesser qualifications (Whalley 1986).

The 1960s were the decade of the white heat of scientific and technological revolution and the overriding ethos behind the expansion of jobs for graduates was the perceived need for major research and development programmes. To some extent the relatively distinct groups of graduate scientists and engineers shared a common ideology, that the universities would produce more of the basic science and the labour power, that the public labs would do some of the basic research and development of new products and processes, and would assist industry with the testing and control of their more advanced industrial processes, while the corporate labs would mix research and industrial development and the production and industrial managers would concentrate on getting goods out of the factory gates. It was, precisely, the development on a larger and larger scale of the linear model; and the separation of stages, particularly marked as we have seen in the UK, involved also the evolution of an increasing, and hierarchical, division of labour. If it was a collective effort in some senses, it also involved an increasing differentiation between groups of workers.

Moreover, the differentiation between management (dimension A) and research and scientific workers (dimension C) also continued and deepened. While 'going into management' might mean promotion it would also probably mean leaving behind the work one was interested in, so relatively few engineers crossed over. Moreover, on the management side in industrial companies the hierarchy was related to managerial position and experience rather than to any educational (and certainly not scientific) qualifications. The engineer who became a production manager would be working in a culture which, as often as not, took the term 'theoretical' as one of abuse (Whalley 1986: 58). This was an attitude found by Whalley to exist also even within the ranks of engineers and scientists. Most prominent in the 'lower ranks' (read: closer to production), where the emphasis was on the need for 'practical men' (*sic*) and the 'real world' of the factory as opposed to the 'airy-fairy ideas' of the classroom, it was still evident in anti-academic rhetoric even within high-technology research groups (Whalley: 58). This was, and still is in many traditional companies, so much the case that there was much cynicism about the usefulness of high-powered graduates, except in R&D labs, and even there the term 'boffin' or 'intellectual' was one of ridicule or denigration. What all this underscores is the particularity of the British social structure in this area. While within industry the gap between mental and manual might be blurred, even within many of the R&D laboratories (which were typically more D than R anyway) the gap between industry and academe was far wider. And that gap was evidenced not only in the technological development (or lack of it) in industry but also in social attitudes and the day-to-day culture of the workplace.

Debate

The class map produced by Wright in his early work is, precisely, a picture of locations. For that reason it is static, in the sense that it says nothing about the trajectories – or indeed the characteristics – of the people holding positions within it, other than their locations on this framework. This is a point which is particularly relevant to any consideration of the social position of scientists and engineers in the UK. For Whalley and Crawford (1984), for instance, labour-market position, and in particular career structure, is an important component of class position. 'For technical workers ... , the total range of positions they can normally expect to occupy during their working lives is more important than temporary occupancy of particular jobs, and a different structuring of careers is a significant component of national variations in the social organization of technical work' (239–40).

The historic importance of apprenticeships in the British case illustrates this point forcibly. Moreover, the issues it brings to the fore are central to our discussion here. It has been argued that the early importance of working-class mechanics reinforced the subsequent reliance on apprenticeships, on-the-job training and promotion from within (Whalley and Crawford 1984: 246), because of the consequent 'unwillingness of the English middle class to participate in technical activities after the mid-nineteenth century'. Distancing from production as an element of status in this field is thus already apparent a century and a half ago. The existing social structure and the evolving models of science and innovation were mutually constitutive. Moreover, the importance of the apprenticeship system had other effects. It served to 'reinforce the general feeling that engineers were not suited to the higher ranks of management' (Whalley and Crawford 1984: 246) and was thus part of the basis for the small number of scientists and engineers in the UK on dimension A. It also influenced the social character of that dimension, therefore: there is less relation between external educational qualifications and managerial level in the UK than is normal in, say, France. The significance of the apprenticeship system also worked more generally to hold down the status of 'engineers' in the UK. In France the distinction between *cadres* and *techniciens* has been more marked and is far more difficult to cross. *Techniciens* have formal, externally awarded credentials and higher status than those holding the same position on Wright's map in the UK. In the UK the maintenance of the apprenticeship system combined with the middle-class disdain of physical production reinforced the feeling that engineers 'were significantly inferior to more respectable professions such as accountancy' (Whalley and Crawford 1984: 246). There was, in other words, a process of mutual influence between the formal class map of Wright, and its real character, the model of science and innovation, and already constituted class cultures and social attitudes which drew on wider forces in civil society than those of the immediate production process. As we shall see, this is a mutual influence which is still crucial today.

Looked at in the other way, from the point of view of individuals located, and maybe moving, within the class map, the British apprenticeship system either provided or blocked opportunities for advancement. It had its own ambiguities. On the one hand, it certainly enabled some working-class people to get a good technical training and to have the potential for moving at least some way up the hierarchies of management (A) and scientific worker (C). Non-academically-credentialled shop-floor workers were provided with an opportunity for advancement within the workplace. When they emerged from the apprenticeship training they had the status of skilled (manual) workers. They were strongly demarcated from semi-skilled workers, they would quickly be responsible for some apprentices in their turn, they had a degree of autonomy in their labour process and, as we have seen, the potential for further advance. To the extent that it was this system of social mobility which lowered the more general status of engineers in the UK, calls for an increase in their status have thus to be viewed with circumspection. On the other hand, while this system provided a bridge for some, it did not alter the fact of a hierarchy (though it changed its nature somewhat, as we have seen); nor did it – nor could it – provide a bridge for everyone. Tight forms of social closure surrounded entry into the system. 'Respectable' families' children got their tickets more easily, especially if they were from the 'right estate', if previous generations had worked in the firm, and if (it hardly needed saying) they were male and white.

In 1985 Wright published his own critique of his previous work and developed an alternative framework. Central to his dissatisfaction with his previous approach was the conceptualisation of the dimension on which the majority of scientists and technologists in the UK are located – that which relates to individual autonomy in the work process. This dimension was criticised not for empirical incorrectness (and we saw in chapter 4 that a relatively high degree of autonomy did characterise this type of work at least on science parks) but for its conceptual inadequacy. Wright (1985) developed a number of 'conceptual constraints' which, he argued, any adequate concept of class must accord with. Among these were: that the concept of class is a relational concept; that the social relations which define classes are intrinsically antagonistic rather than symmetrical; and that the objective basis of these antagonistic interests is exploitation. The concept of autonomy, he argues, satisfies none of these criteria; it is a descriptive gradation rather than a relational concept in which the defining differences are intrinsically antagonistic in the sense of exploitation. One person's autonomy does not necessitate a lack of autonomy for another. Given the results of our enquiries in chapter 4, it is apparent that this is an argument which needs to be put with care. For it was clear there that the *kind* of autonomy we discovered on science parks and the social relations on which it depends is not generalisable to the whole of the population. Indeed, we made this argument specifically at the end of the last chapter. The fact that these jobs require others to service those who do them, and to fit into their timetable, is a case in point. In this kind of arrangement, one person's flexibility is another person's constraint. In this social context, autonomy is not simply a descriptive gradation. But, as the quotation from Elson in the previous chapter pointed out, this is the *class* answer to the need for initiative within the workforce. There could be other answers in which, in principle, all workers could have a higher degree of autonomy. The status of autonomy as relational or not thus depends empirically on the concrete social situation in which it is embedded – in other words, on its concrete meaning, However, there were other reasons, too, why autonomy as a class criterion as characterised by Wright was inadequate. It does not in fact have the petty-bourgeois character he attributed to it, it is structurally underdetermined, and in practice it ran into numerous empirical anomalies. But the fundamental problem, as seen by Wright, was its lack of basis in a relational, antagonistic and exploitative social structure.

To remedy these defects Wright (1985) therefore adopted a new

Figure 5.2 Wright's (1985) typology of class locations in capitalist society

Assets in the means of production

	Owners of means of production	*Non-owners (wage labourers)*			*Organisation assets*
Owns sufficient capital to hire workers and not work	Bourgeoisie	Expert managers	Semi credentialled managers	Uncredentialled managers	+
Owns sufficent capital to hire workers but must work	Small employers	Expert supervisors	Semi credentialled supervisors	Uncredentialled supervisors	>0
Owns sufficient capital to work for self but not to hire workers	Petty bourgeoisie	Expert non-managers	Semi credentialled workers	Proletarians	–
		+	>0	–	
			Skill/credential assets		

Source: based on Wright (1985: 88)

framework (figure 5.2). This time it was based centrally on the ownership/non-ownership of assets, of which three types were identified. First there is a fundamental distinction between ownership and non-ownership of assets in the means of production. Second there are organisational assets. Third, there are skill/credential assets. It is the dimension of skill/credential assets which is most relevant to the discussion here and, interestingly, it is also this characteristic which poses Wright most problems.

First of all, it is difficult to measure 'skill' and in the end, as the nomenclature on the diagram indicates, much of the empirical work is focused on formal credentials, along with occupational titles and job traits.

Second, however, this precisely indicates a further problem, since as we have seen in the case of the UK people with very different formal credentials who can have the same occupational titles and be working in the same jobs. People with academic qualifications, those with apprenticeship qualifications, and even some with long years of on-the-job experience could be working in similar positions. Quite how the bundle of three characteristics would be balanced to achieve a final classification is unclear.

Third, moreover, the tendency to replace the concept of skill with that of credentials is reinforced by the fact that the latter is easier to handle conceptually, given the conceptual constraints which Wright employs. The centrepiece of these constraints, and the main argument of the later book, is as we have seen that while degrees of autonomy and other criteria employed by Wright in his earlier work captured aspects of inequality and of domination they did not all relate unequivocally to exploitation. And it is this relation which Wright sees, in our view correctly, as fundamental to any notion of class. Exploitation is in turn defined relationally: one class in an exploitative relation appropriates the fruits of the labour of the other class, and the two sides of this antagonism are tied together by mutual dependence. 'In the case of economic oppression, the oppressors' material interests would not be hurt if all of the oppressed simply disappeared or died. In the case of exploitation, on the other hand, the exploiting class needs the exploited class' (Wright 1985: 75). The appropriation of labour occurs through the inequality of claims upon the total social product: 'To appropriate the fruits of someone else's labour is equivalent to saying that a person consumes more than they produce' (p. 75). The problem is how one defines all this in relation to credentials and skills. As far as credentials are concerned, Wright sees the appropriation of labour occurring through the fact that credentialism precisely allows, through social closure, restrictions on the supply of the relevant labour power. This in turn allows the wages of those workers to be driven up beyond the cost of the production of that labour power, which Wright argues would be its 'just' reward. Whatever one makes of this argument for expropriation through credentialism, the problems are far greater when it comes to considering skills without formal credentials. At this point Wright resorts to a notion of 'talent', and clearly realises that the ground is somewhat shaky beneath his feet. 'While I cannot give a rigorous defence of this position, I think that it is appropriate to regard the extra income that accrues to people with talents (i.e. people who acquire skills through the deployment of their talents) as a kind of "rent", quite parallel to the rent obtained by the owner of a particularly fertile land' (Wright 1985: 76). This is surely problematical.

But it is related to a further issue. For in fact Wright remains dissatisfied with this dimension of his new class framework. Thus he feels it might fail on one of the conceptual constraints: 'it is not at all clear', he writes, 'that one can derive any relational properties from the ownership of skill assets as such'. He continues 'To be sure, if skill assets are a criterion for recruitment into positions within organisational hierarchies, then individuals with skills or credentials may be in a particular relation to people without such credentials, but this is because of the link between skill and organisation assets, not because of skill assets themselves. The most one can say here is that experts and non-experts exist in a kind of diffuse relation of dependence of the latter on the former. This is a considerably weaker sense of "social relation" than is the case for the other three types of class relations' (p. 85). In a footnote he adds 'In the case of capitalist societies this might imply that skill or credential differences should be regarded as the basis for class segments or fractions among workers and among manager-bureaucrats, rather than a proper dimension of the class structure. I will continue in the rest of this book to treat credential-exploitation as the basis of a class relation ... but this characterisation should be treated cautiously' (p. 103).

Now, there are a number of issues here. While it is not our aim to come down on one side or the other of a dichotomised argument about whether skilled scientific workers do or do not constitute a class (if anything it seems really unlikely that they could be said to), we would take issue with aspects of Wright's arguments. First there is the question of relational definition. Wright's concepts of skill and credentials are, as he says himself, not really strictly relational, and this is most certainly true in the case of non-credentialled skills. Yet surely this is inadequate. What is at issue here is not individual talent or innate ability but the social construction of divisions of labour. (Wright's tendency to focus only on the individual and to lose sight of the social whole is related to his reliance on Roemer and game-theoretic procedures.) While the fact that one person can sing well (or is good at designing computer software) does not mean that another is deprived of the possibility of also being so accomplished, society's recognition of differences is what creates the possibility for some to exercise (and charge for) their skills to the exclusion (and this is the point) of others. Even in the case of 'talent', therefore, the reason for the rent at all is not (or not only) innate skill (whatever that is) but (also) the limitation of opportunities to use it. Further, Wright argues in the passage cited above that non-experts are organisationally dependent on experts. This is true. But in the wider context of the division of labour experts are also dependent on non-experts. The categories *are* in this sense relational. Moreover they are also mutually defining in the sense that the existence of 'experts' implies necessarily the existence of 'non-experts'. Thus, in the case of the linear model (see figure 3.1), the fact that there is a separate activity 'applied research' implies also that there must be upstream activities (basic research) and downstream activities (experimental development, full production, etc.). Moreover, the existence and nature of, say, activities 1, 2, 4 and 5 determines both the existence and the nature of activity 3, and the element in the division of labour attached to it. The elements are mutually constitutive.

Further, in a general sense, rather than in the case of specific expertises (not all of us can do our own plumbing, mend our own machines, build our own power-generators), where the notion of credentials and skill is heavily reliant on the separation of conception from execution, as it is in our societies, and most particularly in the case of the linear model of science and innovation, then surely the relationship is, in a general sense if not precisely in Wright's, antagonistic.

But it is so in a way which challenges the deeper roots of Wright's argument. Here we return to his point that exploitation must involve the appropriation by some of the fruits of the labour of others. It is a definition of exploitation which is essentially rooted in the *distribution* of income (as, indeed, some versions of profits and wages are). Becker (1989) has also commented on this. It is not necessarily wrong, but it may be too limited. There is more to exploitation than that. First, the separation of conception from execution gives the power of strategic control and decision-making within the system unambiguously to one side, and at the expense of the other. (Similarly in relation to capital and labour it is not only the amount of profit appropriated but the control by the capitalist over the criteria of investment and the direction of the economy which are at issue in class relations.) Second, skills/credentials are surely not just an asset because they bring in income. The development of skills and talents is a positive good in itself. The problem is that, while it is controlled and shaped by the demands of the present system of production, and in particular the division of labour characteristic of the UK, it remains a highly unequally distributed good. We hark back here again to the arguments of Elson rehearsed in chapter 4: while in principle the development of skills by one person does not impede their development by another, in practice the requirements of the labour market may in fact introduce impediments. Moreover, while in the long run it is surely true that the free development of each depends on the free development of all, in the short run – as Wright says – those with education and training benefit from their scarcity value. (This differentiation in relation to time period raises, of course, the problematic notion of 'interests'.) Finally, what is at issue is not the problem of the individual credentials and talents which Wright is wrestling with, but the social opportunities to develop and use them.[1] Indeed, any notion of skill which is based as much on individual talent as on the demands of the workplace fails to live up to a further conceptual constraint to which Wright himself demands adherence: that the fundamental basis of exploitation is to be found in the social relations of production.

For us, then, skill in the sense in which it is defined in the workplace *is* a relational concept. The designation of some jobs as skilled in particular ways means that other jobs will not need (and other people cannot use in their employment) those skills. In principle a co-operative division of labour with different people developing different skills, but without a relation of hierarchy between them (though there will be one of mutual dependence) should be possible. In practice what we are more frequently faced with, and most especially so in the case of the linear, hierarchical model of science and innovation, is a particular form of the division of labour where the possession of increased skills by some only even further deprives others of the ability to develop (and to gain income from) their own potential. In their physical crystallisation of the linear model, science parks epitomise and – we shall argue – reinforce precisely this.

RECENT CHANGES IN THE SOCIAL LOCATION OF SCIENTISTS AND TECHNOLOGISTS

Traditionally high-status jobs

Among the many changes which have occurred since the 1960s in the broad profile of employment of scientists and technologists in the UK one of the most marked has been a decline in what used to be considered the high-status occupations. This decline has been multi-faceted. The aspect most easily registered is that there has been a simple decline in employment opportunities: a fall in (or in some cases a cessation of the growth in) the sheer number of jobs available. But it has also been more than this. The same political and ideological forces which have led to this numerical decline have also been behind a clear reduction of pay and working conditions, and also of status, attached to the jobs which used to be considered the élite positions for British scientists and engineers. Meanwhile other occupations have been dramatically increasing their financial rewards and their symbolic cachet.

Perhaps most obviously, the traditional opportunities in higher education (teaching and research) and in government laboratories have declined. Tenured or permanent employment in universities has declined. Numbers of wholly university-financed full-time academic staff (as opposed to 'externally funded' mostly short-term contract researchers) dropped from 33,695 in 1978–9 to 30,621 in 1988–9 (UFC 1990). Of these, the numbers in science and engineering dropped from 12,503 to 11,521. The number of new full-time university staff appointed who were below the age of 40 plummeted to around 300 in 1985 and 1986, less than 1 per cent of the tenured staff complement. In 1988–9 the intake of younger academics (1,052) was still just 450 ahead of outflow (at 602). At the same time non-wholly university-funded staff and part-time academic numbers more than doubled (from 8,516 to 17,477) in the decade to 1988 (AUT 1990).

Jobs in government research laboratories and research council labs have also declined, as the *Annual Review of Government Funded Research and Development* indicates. Overall, the 1989 *Review* not only marked past declines, in real terms, it also projected decline into the future. Thus, in 1987/8 pounds, government spending was projected to fall from £4,816 million in 1986/7 to £4,282 million by 1991/2. This is a 10 per cent cut in real terms, and involves reduced spending on almost all categories of government R&D (Ince 1989). Thus jobs in UKAEA, which had employed half of all those in government research labs, have fallen significantly. The Agriculture and Food Research Council stated in the *Review* in 1986 that 'the large restructuring ... is leading to the loss of numerous staff'. The SERC reported significant decreases. The DTI showed stagnant employment and difficulties in recruiting, with 40 per cent of new entrants leaving within three years. In October 1990 the Science and Engineering Research Council reported on a major comparative study of research expenditure in the UK, France and (West) Germany. It was the first study to compare spending at normal currency exchange rates rather than using purchasing power parities. On public research expenditure it concluded that the level in the UK was £4.1 billion, compared with £6.8 billion in France and £7.2 billion in West Germany, and that government *civil* R&D budgets were three times as high in Germany, and twice as high in France, as in the UK. The report also pinpointed a lack of support in the UK for 'basic and strategic' research, being about half that in West Germany and two-thirds that in France (Williams 1990). All this has serious implications for the career prospects of today's younger researchers. In universities, for instance, until the 1970s the classic trajectory would have been for postgraduates staying in academe to move from short-term contracts to permanent lectureships. 'Today the total number of new university lectureships comes nowhere near to meeting the demand from contract researchers at one university the size of Edinburgh, let alone for the whole of Britain' (Smith, T. 1987).

Alongside this decline in the number of opportunities, moreover, salaries in the universities and the public sector more generally have fallen significantly behind those for equivalent jobs in some other parts of the economy. Even as early as the mid-

1980s, the DTI was blaming its recruitment difficulties on civil service pay rates. Conditions of work also fare badly in international comparisons. Thus the SERC report cited above found that in basic and strategic research in the UK each researcher had £53,000 spent on them while the figure in West Germany was £93,000 (Williams 1990). Moreover, the attractiveness of such jobs to all but the most committed high-fliers must have been severely reduced by the ideological opprobrium attached in the 1980s to anything to do with 'the public sector', while the term 'academic' is now in society more generally a term of comic abuse.

None the less, some of the intellectually-inclined graduates have not reacted by changing their career preference. Rather they have continued their academic career abroad by joining the worldwide movement of scientists from cheap-location countries to countries where institutions can pay more and offer more stable and satisfying conditions. An international division of labour is being reinforced – and changed in its nature – with major implications for the UK's basic science base. The Royal Society study on the 'brain drain' (Royal Society 1987), although showing relatively small emigration rates, demonstrated that the most serious emigration was of recent PhD scientists of high quality from universities. Significant numbers of very high-calibre scientists (including an increasing number of Fellows of the Royal Society) are emigrating, as well as post-doctoral research fellows. Most emigration is from universities rather than other research labs and industry. Moreover, although the numbers are roughly balanced by scientists arriving in the UK, few of the arrivals come to work in this country for longer than three years, many coming for only a few months. In contrast, the majority of Britons who left the UK took up long-term posts abroad.

Lack of more permanent posts in British universities would seem to be an important factor, since almost three-quarters of all emigrants from the UK were in posts of three years or less before leaving. The most commonly cited reasons for leaving were 'career opportunities abroad' and 'career limitations in UK' (joint top) followed by 'rates of pay'. The *New Scientist* cites the report as follows:

> 'The general lack of career structure in university research was severely criticised by a number of respondents – short fixed-term contracts were said to be unattractive to young

> scientists, who sought, if not tenured posts, then at least ones lasting substantially longer than three years.'
>
> Finding replacements was most difficult, among the surveyed disciplines, in electronic engineering. One head of department told the society: 'The problem in electronics is now how to find anybody willing to work in universities. We used to rely on overseas PhDs, but this supply is drying up as other countries offer more attractive prospects. The total number of British electronics graduates is far too few to satisfy demand. The brain drain is only a small influence on the total problem.'
>
> The [Royal] Society agrees. 'Emigration is widely regarded as one symptom of an overstressed system and it is felt to be having an adverse effect on British research when added to other factors such as reduced UK funding for research, increased investment in research in other countries, widening differentials between British industrial salaries and a perceived shortage of scientists and engineers.'
>
> (*New Scientist,* 2 July 1987, citing Royal Society 1987)

The argument which is used to justify cuts in basic science is, as we have seen, often that the UK's problem is that it has lots of Nobel prize-winners but not enough commercially successful innovations, and that one solution is to 'encourage' scientists to switch to more commercially applicable work. Indeed, as we have seen earlier, science parks are seen as agents of change towards commercialising academe. As Lord Young said in his address to the 1985 UK Science Parks Conference:

> We have a long fine tradition in this country of excellent academic research – the number of Nobel prizes awarded to British academics speaks for itself. But that excellence has failed to translate itself into excellence in the market place.

The danger is, however, that current strategies might cut into the areas of strength (there is strong evidence that the capacity to produce Nobel prize-winners in the UK is declining rapidly) while not increasing the move of scientists towards careers in commercialising their knowledge. In part, of course, this reflects that deeper antipathy, already discussed, towards industry and production.

Another reflection of that antipathy, as well as of the changing ideological and political climate more widely and, more simply, of shifts in relative salaries between sectors of the economy, has been the move of graduates in science towards the financial sector and most particularly the City. It is notable that this move is far less marked among higher-degree graduates than among those with only a first degree. None the less, in 1987–8 10.8 per cent of male and 7.5 per cent of female higher-degree graduates in pure science took up their first permanent employment in the financial and commercial sectors of the economy (Association of Graduate Careers Advisory Services 1988). Among first-degree graduates the figures are startling. Table 5.1 provides a rough comparison for the years 1980-1 and 1988–9, and registers a dramatic shift in balance away from other sectors and towards finance and commerce.

Table 5.1 Employer category of first-degree science graduates entering permanent employment in the UK (%)

	Category	*Public service*	*Education*	*Industry*	*Commerce and finance*	*Misc.*
	Men	13.5	6.5	48.9	27.8	3.3
1980–1	Women	25.7	9.5	31.6	29.1	4.2
	Total	17.2	7.5	43.6	28.2	3.6
	Men	11.5	4.0	39.3	39.4	5.8
1988–9	Women	18.9	7.2	30.6	37.5	5.8
	Total	14.1	5.2	36.2	38.7	5.8

Source: UGC (1982); UFC (1990). The figures are for 'biological and physical sciences'

These moves are part of much wider recent shifts by science and engineering graduates into non-technical jobs. Silberston, in his 1987 lecture 'Is there a shortage of engineers?' (a different category from the pure science graduates examined above), estimated that 11 per cent of engineering graduates went into non-technical jobs in 1985, in comparison with 6 per cent in 1982. The figure was 19 per cent from Cambridge, the highest from any university (Silberston 1987: 11). He suggests that the proportion rose again in 1986 and 1987. Silberston compared the starting salary for Imperial College engineering graduates in 1986 (£9,300 to £10,500) with £18,000 for new graduates in the City. The Institute of Physics described the changes as follows:

> Before the very recent popularity of denationalisation and (against) public ownership it was probably reasonable to claim, on the basis of annual salary surveys of IOP membership, that one-third of all UK physics graduates were engaged in industrial research, development or management in private sector industries, another third were teachers, lecturers or researchers within the 'education' sectors, and the remaining third were employed directly or indirectly by government organisations or agencies, mainly in research and development in scientific and defence work. However, the most recent statistics on first destinations show that of the 1097 first degree graduates entering home employment in September 1986, 171 (16%) went into public service, 44 (4%) went into education, 96 into accountancy, 48 into banking and finance, and 125 into commerce and commercial services. In simple terms, 25% chose financial or commercial careers of some kind.
>
> (Davies 1988)

It is interesting to analyse what this attack on the traditionally high-status jobs means in terms of the dominant British model of science and innovation. Most simply, it means a lowering of the priority given to, and the status, conditions, etc. within, the activities in box 1 in figure 3.1. The aim, as we have seen, seems to have been to push people 'downstream' into more immediately commercial activities. The evidence so far, to be reinforced below, is that the strategy has not worked. In part this is because of the wider and long-standing antipathy to production. But in part it is also because simply attacking one element in the linear model will not increase its effectiveness. To bring about the improvement in the links which is desired it would be better to change, or at least to modify, the model itself. But at minimum this would imply improvements in research facilities and funding, and skill levels, within industry itself. And this has not happened.

Traditionally, as we have seen, those relatively high-flying graduates who were interested in more applied research and development took up employment in industrial R&D laboratories. But even here there has been evidence over the medium term of declining or stagnating work opportunities. More recently there have been some indications of increases, but still nothing like as much as in most other European countries, the US and Japan.

Moreover, such spending increases vary dramatically between sectors. Electronics R&D spending more than doubled over the ten years to 1985, corresponding to the growth of the sector, but still compares badly with other Western countries. And even these recent spending increases are partly accounted for by the rapid build-up of R&D by foreign-owned companies, particularly IBM. Chemical and pharmaceutical R&D also increased, but most others stagnated or declined. Mechanical, electrical and aerospace engineering all cut their real R&D spending between 1981 and 1985. The 1990 SERC research found that West German industry funds two and a half times as much research and development as British industry, and that France funds slightly more than the UK (Williams 1990).

Moreover, lower R&D employment should be seen in the context of the already relatively low numbers of scientists and technologists in R&D in the UK. Japan employs four scientists and technologists in R&D for every one in the UK (Rudge 1986). The proportion of graduate scientists and technologists engaged in R&D is extremely low. One in three of non-university scientists and technologists is engaged in R&D in Japan, compared to one in four in the US, one in five in West Germany and France, and only one in eight in the UK. There are clear indications that rationalisation of private and public companies, and privatisation of nationalised industries, has lowered the number of R&D jobs. Senker makes the point that the shortages would be much greater if Britain was not losing international competitiveness.

> There are serious shortages of scientists and engineers in Britain, significant only in relation to the objective of enhancing British industry's international competitiveness. Broadly speaking, supplies of scientists and engineers are generally adequate to sustain the present trend of Britain's slowly declining international competitiveness in markets for manufactured products.
>
> (Senker 1991)

And career opportunities in R&D may also be diminishing. MSL International's survey of job advertisements in prominent journals reports a strong downward trend in jobs advertised for R&D managers, the higher-level R&D jobs (Dixon 1988). In 1987 posts advertised dropped by a further 8.4 per cent, on top of a dramatic fall of 44.8 per cent in 1986. Thus in 1987 the number of R&D

management posts advertised, at 3,374, was only half those advertised in 1985. After a rise from 1987 to 1989 a further sharp decline left the level at 3,273 for 1989–90 (Dixon 1990).

Finally, and paralleling the situation in the university sector, the pay and conditions in research and development posts lag behind those in other, otherwise comparable, jobs. The 1990 SERC research found levels of expenditure for each researcher in British industry only half those in West German or French industry, and reported that each British industrial researcher has about 1.1 support staff, compared with more than 1.9 in France and West Germany (Williams 1990). Thus an Incomes Data Services (IDS) survey in 1987 showed research and development staff to be undervalued by most companies in the salaries they received, particularly compared with managers. It found a symptom similar to that explored in an earlier section – that R&D specialists hit a 'salary bar' (in 1987 between £16,000 and £20,000 p.a.) beyond which they could not progress unless they switched to management or to other specialist areas. Incomes Data Services commented that UK salaries were low compared with those in the US, 'where many of the estimated 1,000 emigrating British R&D technologists are being lured every year' (Brindle 1987; IDS 1987). A report in 1988 indicated that 'growing staff shortages' were beginning to push up the wages of research workers (Reward 1988; *Financial Times*, 23 August 1988); but it remained isolated evidence. Moreover, again in 1987, a detailed cross-European survey put these results in another international perspective. Table 5.2 reproduces the results, which compare the salaries of directors of different departments. The UK shared with Portugal, Greece and Spain the characteristic of paying the finance directors the highest salary. And it shared with Belgium and Spain the characteristic of paying research directors least. As the commentary in the *Financial Times* pointed out,

> the British companies seem to value them by far the least of anyone. The Belgian research heads' total money rewards were 2.9 per cent lower than the pay of the country's departmental directors as a whole, and those of the Spanish were 7.1 per cent lower. By the same measure their UK counterparts were paid 11.2 per cent less.
>
> (Dixon 1987)

A new hierarchy?

But if the trend of recent years among the traditionally high-status jobs has been predominantly one of decline, the changes elsewhere among practising scientists and engineers have been a lot more complex.

It has already been seen that work in more traditional engineering companies has historically not been viewed very positively. There have been numerous studies showing that it is the relatively less qualified graduates who go into production jobs. The status of industrial scientists and technologists in production has been perceived as very low.

There is no evidence to suggest that the poor image of engineering and the problems of career progression in traditional industrial production have improved in recent years. Whalley's (1986) study is a case in point, and there are numerous surveys of engineers bearing witness to their dissatisfaction. In 1989 a survey of UK engineering graduates revealed that two out of three emerging on to the job market that year had decided against a career in engineering. This included those who had been turned off an engineering career by their thirteen months of work experience in companies – nearly half those who had had such experience. Adjectives such as 'downright boring' and 'past it' (applied to the companies, their managers and their promotional literature) combined with perceptions that engineers were not given enough interesting or important projects to handle, and an awareness of the low level of pay (IVL 1989; Cornelius 1989; Garnett 1989). Later that same year, a survey conducted by *The Engineer* of people actually working in engineering came up with similar results. This time not only university graduates but also those with HND and HNC qualifications were surveyed.

> Over 90% feel engineers are underpaid and undervalued, and given a second chance only 36% would repeat their careers in engineering ... most would seriously opt for a career in another profession. Nearly a quarter would become accountants, 14% general managers and 7% would go into the media.
>
> (Eustace 1989)

But conditions in some parts of manufacturing industry, and its related activities, *have* been changing. Even the surveys we have just cited showed some careful distinctions being made. The IVL

Table 5.2 International comparison of salaries of departmental directors

Country	Average total pay of chief execs. £	Average total money rewards of departmental directors as percentage of chief executives' average							
		Finance %	Marketing %	Sales %	Personnel %	Production %	Research %	Engineering %	All departments %
Portugal	19,154	73.1*	72.5	72.2	72.2	70.1	69.5	63.7†	70.5
Irish Republic	46,047	69.5	64.3†	72.9*	71.3	72.7	-	69.2	70.0
Netherlands	63,285	67.7	72.3	73.9*	61.9†	68.8	73.2	67.7	69.3
Norway	48,700	65.2	68.1	78.0*	69.4	65.8	-	62.2†	68.1
Greece	22,522	76.8*	70.5	70.0	62.2	73.2	-	54.7†	67.9
Belgium	69,377	65.4	67.6*	65.6	65.2	65.9	63.3†	63.3	65.2
West Germany	90,655	62.4	65.5	66.2	60.0†	64.5	69.1*	60.9	64.1
Spain	52,647	68.4*	62.7	62.7	58.7	59.5	57.4†	62.8	61.8
Sweden	55,969	62.5	62.9*	62.5	60.6	62.6	-	59.8†	61.8
Italy	71,183	63.6	61.6	62.0	53.2†	64.8*	60.4	60.1	60.8
France	73,492	62.0	62.9*	61.9	57.8	58.7	58.6	57.6†	59.9
UK	53,730	65.8*	64.7	65.2	57.3	54.3	53.2†	58.5	59.9
Switzerland	109,412	57.7	65.0*	60.3	63.0	53.4	56.2	49.0†	57.8

* Highest paid departmental director. † Lowest paid departmental director
Source: Dixon (1987)

survey found that the best-received firms were in the 'information technology' and 'electronics' sectors; the worst reports were on traditional engineering sectors and construction (Cornelius 1989). And it seems more generally true that scientists and engineers working in the electronics and related sectors can lay claim to a status not easily available to those working in the rest of manufacturing. At A level, even, the top entry grades to university are in electronic engineering, followed by civil engineering, rather than, for instance, in mechanical engineering.

There are a number of reasons why this is so. To some extent it is the simple fact of electronics and information technology being relatively new, with all the hype of high-tech attached to them. It is also that there is more higher-status work in these sectors. A 1986 survey by the Engineering Industry Training Board (EITB) showed that within manufacturing the biggest increases between 1978 and 1985 in the employment of professional engineers, scientists and technologists had come in electronics and in office and data-processing equipment. These two sectors accounted for two-thirds of the increase in employment. Overall, one in ten of total employees in electronics, and one in six in data-processing equipment, were scientists and technologists. There was also a far higher proportion of technicians – one in seven – than in other sectors. These things give an industry, even one within manufacturing, a different image. Moreover that image of higher status for engineers is further enhanced in the eyes of many by the fact that there is in these sectors a sharper division between mental and manual labour. The intellectual labour in electronics seems more purely intellectual, and is certainly more removed from the necessity to be on the shop floor. The distance from production, even here within manufacturing, is increased. That pushing forward of the division between conception and execution, moreover, has taken place against the background of a wider shift away from on-the-job training and apprenticeships and towards formal and more academic credentials. This, in turn, has been a response to differential sectoral performance and – most particularly – to political policy. At craft level, there has been a concerted effort by the Conservative government and some employers to break the apprentice craft system. Attempts have been made both to merge previously different crafts, like mechanical and electrical, but more significantly to break the divide between certificated craft occupations and work designated as semi-skilled or unskilled. The number of craft workers in engineering industries fell by 180,000 to 352,096 from 1978 to 1985. And between 1980 and 1984 recruitment of craft and technician trainees in general by industry fell by 50 per cent (Taylor 1987). The number of craft trainees in EITB industries completing their first-year basic training fell from 16,985 to 5,400 in the ten years to 1988. Similarly, technician numbers fell from 6,022 to 2,873 (EITB 1989: 50). This shift towards academic credentialism has been a major feature of the changing profile of science and engineering occupations. And it presumably reinforces that other apparent attraction of electronics, its comparative cleanliness. Instead of the physicality and the grime of mechanical engineering, here the nearest one may have to get to a machine is a computer on the desk.

Moreover, in the case of information technology, closely related to electronics, a major portion of the work is in software. And much of software production is classified not as manufacturing but as 'services', which further shifts the imagery. Software production grew with a major division between 'craft' programmers, most without higher educational qualifications, and systems analysts, a group which included more graduates. Both groups are 'mental' workers with relatively high status, and the division between them can be fuzzy, with 'anarchic' career structures (Ince 1988).

It may also be that the status of scientists and engineers in these sectors has been further enhanced by a completely different rhetoric. There has been an increasing emphasis on the virtues of employment in small-to-medium-sized firms and the development of an ideology of scientific and technological entrepreneurship: the view that technologists are better-off running new risk-taking companies than remaining in safe and frumpish larger companies: 'good salaries and pension rights at British Telecom encourage people to stay when many probably have the ability to start their own businesses' (Walker, cited in Marsh 1987).

As we have seen, this is a novel situation in the UK. There is no recent history in the UK of significant numbers of scientists and engineers starting their own businesses. Indeed, the lack of opportunity for self-employment of engineers has been given as one reason for their weak professional status, in strong contrast to the situation in Germany and France. Unlike lawyers, doctors, dentists and architects, engineers have been predominantly employees rather than independent practitioners.

> They have had to sell their expertise to powerful corporate clients who reserve for themselves the right to judge an acceptable performance, and their close association with manufacture has sometimes produced status problems in societies where making things is the preserve of the working class.
>
> (Whalley 1986: 5)

Thus, since the second world war, both scientists and engineers have tended to work as employees in large organisations: the universities, central government and private R&D labs, and big companies. Even when work autonomy for individual staff has been relatively high, such organisations have tended to have a 'traditional' very hierarchical structure, with long career ladders, and with seniority playing a big role compared with skill and training.

There is some evidence that recently there has been an increase in self-employment and 'entrepreneurship' among scientists and engineers. Savage *et al.* (1988), for instance, cite indications from their work in Berkshire. More broadly, a 1987 Engineering Council survey of members found 6.8 per cent of chartered engineers who responded to be self-employed. This compared with figures of 2.7 per cent in 1966, 4.1 per cent in 1983 and 6.6 per cent in 1985. The figure for 'consultants' had also increased, from 8.5 per cent in 1983 to 10.0 per cent in 1987. The picture was the same for technician engineers, the figure for the self-employed having increased from 3.0 per cent in 1977 to 5.6 per cent in 1987 and for consultants from 2.5 per cent to 4.0 per cent. These are not major changes but they are notable, and they do of course fit precisely with that process of scientist/engineer-turning-entrepreneur which science parks are aimed at encouraging. Moreover, the electronics industry and associated sectors have more widely been associated with the imagery, as we saw in chapter 4, of small firms and entrepreneurship.

The evidence on size of firm is interesting. The EITB (1986) survey showed that the most important sectors for the employment of professional engineers, scientists and technologists were electronics, aerospace, office and data-processing equipment, and machinery. The figures for their employment by size of firm are given in table 5.3. It is certainly true that electronics and office and data-processing equipment employ a large number of QSEs in firms with fewer than 100 employees, and it must be on this

Table 5.3 Professional engineers', scientists' and technologists' employment, by selected sector and size of establishment, 1985

Sector	*1-99 employees* No.	%	*100-999* No.	%	*1,000+* No.	%
Aerospace	122	0.9	775	5.7	12,821	93.4
Electronics	1,680	6.9	9,879	40.6	12,759	52.5
Office and d.p.e.	1,062	8.1	6,188	47.2	5,852	44.7
Machinery	1,088	11.9	4,955	55.7	2,858	32.1
Total of all sectors	6,216	7.3	34,104	40.0	45,017	52.8

Source: EITB (1986: table A6)

that the imagery is based. Certainly the contrast with aerospace, also high-tech but a classic industry of the 1960s, is very clear. But in proportional terms the evidence is far from convincing. Over half of professional engineers, scientists and technologists in electronics work in firms with over 1,000 employees, and another 40 per cent in firms with between 100 and 999 workers. The large absolute number working in smaller companies amounts in fact to only 6.9 per cent of the sectoral total; hardly 'representative'. Moreover, if the small-firm imagery is to reside anywhere it should surely be with the machinery sector, which has the second highest absolute total and the highest (among these sectors) percentage. Indeed, if representativeness of sector were really the issue, then small firms (on this definition) are far more important for such employees in machine tools (43.8 per cent), other metal goods (27.2 per cent), metal manufacture (22.6 per cent) and foundries (18.4 per cent). In machine tools *all* such employees worked in firms with fewer than 1,000 workers. But these are sectors which, on other dimensions, are at the opposite end of the spectrum from electronics. All these sectors had less than 1 per cent of professional engineers, scientists and technologists in their total workforces. The imagery of 'metal-bashing' and of foundries is far older, far closer to physical production and – in spite of these figures – far further from the modern image of entrepreneurship and small firms. No one characteristic on its own can account for the higher status of electronics, nor can the characteristics be simply added up. They form a mutually reinforcing articulation, of sometimes unrepresentative images, which together adds up to the symbolism of 'high tech'.

Reflections

In 1988 an article entitled 'Where dishevelled dons are giving way to boffins with more BMWs than bikes' ran:

> Cambridge's traditional quota of dishevelled dons and mad inventors is sadly dwindling. They can still be spotted, pedalling venerable bicycles to gloomy laboratories where research progresses untainted by enterprise culture, but their days are numbered.
>
> Now the ambitious young Cambridge science graduate is cast in a very different mould. He (most are male) arrives at the gleaming buildings mushrooming on the outskirts of the city amid the surrounding villages, immaculately turned out in this season's Next suit and driving a black BMW ...
>
> The centre's [St Johns Innovation Centre] director is Mr Bill Bolton, an ex-Cambridge engineering lecturer with his own small robotics firm and not, he insists, 'a toffee-nosed academic'.
>
> (Buxton 1988)

To summarise the changes, traditionally high-status jobs for scientists and technologists have been cut. In response, some few well qualified graduates have gone elsewhere in search of academic posts and Nobel prizes. Others have swum with the tide towards the City and commerce and seemingly away from technology. More new graduates have gone into private industry in newer sectors. There are some indications of changes in engineering work, though it is not clear that a career in industry is thought of any more highly. There has been strong growth in that sci-tech employment with an image of new high-status work – epitomised precisely by work in science parks.

These changes represent some interesting shifts in terms of Wright's class maps. In terms of the 1978 version the clearest changes have been on dimensions B and C (figure 5.1). The proportion of scientists on dimension B, between petty bourgeoisie and bourgeoisie, seems definitely to have increased, though perhaps not as much as the rhetoric might indicate. On dimension C, the semi-autonomous employees placed by Wright between proletariat and petty bourgeoisie, there have been changes in character. Two forms of dichotomisation have been strengthened. On the one hand, as we have seen, there are the increased distinctions between traditional and high-tech work, with the latter claiming more status and autonomy. On the other hand, the divide between manual and mental labour is increasing. Not only have graduates come into industry to take over more of the higher-status jobs, but the technical division of labour has changed, too, especially in the newer sectors such as electronics and information technology. The skills the apprentice crafts provided have come to be seen, by some, as less needed. And the combination of all these things has meant in the UK that some more of the technical knowledge which used to reside there has been taken away from the shop floor. The position in relation to dimension A, management, is more difficult to pin down. The figures are ambiguous. While considerable numbers of science graduates go into management and management services it is not clear either that they go in *as* scientists and engineers or that scientists and engineers who have worked already in industry cross over any more frequently into management. Indeed, the increasing formal credentialism in management itself is likely to make this process more difficult (as well as making such progression more difficult for the person from the shop floor) – although the qualifications of management, even though increasing, remain low (Institute for Employment Research 1989). There are clearly, as was seen in chapter 4, some increases in participation in management, particularly in smaller firms. And there may be here some differences, within the UK, between British-owned companies and foreign-owned ones. Certainly there is no indication that the overall traditional ethos and attitudes have changed at all.

But it is the debate around Wright's later framework which is more relevant to these recent changes. The rise in importance of formal academic credentials is crucial here. On the one hand, as would be expected, given the discussion in earlier sections of this chapter, it has helped to increase the status of engineering and technical occupations. On the other hand, it has done this alongside *and at the expense of* a further removal of skills from the shop floor, a process reinforced by the dramatic decline of the apprenticeship system, a decline which in itself may further increase the social standing of the credentialled. It seems hard to evade the conclusion that if the status of scientists and engineers, particularly in manufacturing and related industries, has been rising, this has occurred to the benefit of some and to the detriment of others, in terms of location on a status hierarchy, real skill content and chances of individual progression. It is empirical confirmation of the relational nature of 'skill' in social contexts such as these.

In a more general sense, too, recent years have witnessed a process of polarisation. The emergence of a new hierarchy, a new élite, has been associated with a much documented failure in education and training policy which has been exacerbated in the 1980s. At a very local level, in Cambridge, the very firms that are part of the image of 'new' work are reluctant to release their more lowly employees for training:

> They are often reluctant to release key members of staff on training courses for any length of time because of the cost pressures. In-house training is also costly when members of staff and company resources are involved in training.
>
> (Cambridge Local Collaborative Project 1986: 46)

This report says that small firms get their skilled staff by poaching them from larger companies.

More generally, the skill shortages that bid up the status and conditions of one small stratum of employees must be seen in the context of low educational and training levels for other employees. The fraction of élite, skilled jobs requiring scientific, technological and entrepreneurial skills must be counterposed to the extremely weak skills training and educational level of the mass of the British workforce. The contrast between the spending on training of industrial companies in Japan and Britain is enormous. Prais (1981) identified a large proportion of the British labour force who have no vocational qualifications at all – about two-thirds of the workforce, compared with one-third in the then Federal Republic of Germany. Later work suggests that those in the German workforce gaining vocational qualifications did so 'at standards which are generally as high as, and on the whole a little higher than, those attained by the smaller proportion in Britain' (Prais and Wagner 1983: 63). And this was before the really dramatic decline in technician and craft training documented by the EITB (1986).

Finally, at basic educational levels, there is a shortage of science and maths schoolteachers.

> Recent evidence suggests that maths and science teaching in British schools is in a state of crisis. A long standing shortage

> of specialist teachers in these subjects appears to be worsening ... In the case of physics, the DES Assessment of Performance Unit reports a lack of qualified teachers in 70 per cent of junior comprehensive schools and 22 per cent of all secondary schools. In 1985 a survey by the Secondary Heads Association of 831 schools in England and Wales (roughly a fifth of all secondary schools) found that in total they had lost at least 500 maths or science teachers to jobs outside teaching in the previous twelve months. Such losses are compounded by a decline in new recruitment ... The shortcomings of science education in British schools go far beyond specialist teacher shortages. The 1986 conference of the Association of Science Education (ASE) heard reports that many schools cannot afford adequate laboratories, equipment, materials or ancillary staff ... The basic problem is self-reinforcing; inadequate science teaching now feeds into relatively low numbers of science graduates and potential science teachers in the future.
>
> (EITB 1986: 29–30)

There has for centuries, even if sitting uneasily in the UK with the low status of the engineer, been a mystique – something of a fantasy – around science and the current forms of advanced technology. It seems that what we may have been witnessing between the 1960s and the 1980s is some shift in the nature of, and the basis for, that mystique. This is a different bundle of characteristics, built around a greater emphasis on the private sector than the public, on entrepreneurship rather than the big laboratory. As we have seen, both in chapter 3 and in this chapter, not all the rhetoric is backed up by a close inspection of the data. At one level, however, this does not matter; imagery can have its own effects, and the imagery here is contributing to the boost in status for the new breed of scientist and technologist. There have been other shifts, too. From the corporatism of the 1960s to a greater emphasis on individualism, even from a reasonably high level of trade unionism to a lower one today. Unionisation of such 'white-collar' staff gradually increased, particularly in the 1970s, mostly in public-sector unions like Nalgo and the IPCS, but also in ASTMS and TASS. An Engineering Council survey of members found that 41 per cent of chartered engineers and 54 per cent of registered technicians belonged to a union in 1985. Already by 1987 this had fallen to 36.7 per cent (Engineering Council 1985, 1987). The growth of unionism and the notion of modernisation particularly linked with the 1964 Wilson Labour election victory on the call to 'the white heat of the scientific and technological revolution' was a culmination of the view that such 'middle class' groups could have interests in common with the Labour Party. But some shift had taken place by the end of the 1970s: 'The Wilsonian seizure of science as the theme for the 1964 election ... sought to define and capture a new stratum – that of scientists and technologists – and to insist that their interests lay with Labour ... A high proportion of scientists and technologists must have voted Labour in 1964. Almost certainly, a high proportion of the same group voted Thatcher in 1979' (Rose and Rose 1982).

It is not so clear where political allegiances would lie today. But that change is crucial to an understanding of the current high-tech fantasy: the move from the idea of sci-tech advance as coming from the big labs of corporations fed by state support in basic and defence science to the idea of scientists and technologists as entrepreneurs in small innovative companies that are at the same time near to the market. It is a far cry from the classic post-second world war idea of industrial science and technology.

The evidence cited in chapter 4 made it clear that, on science parks at least, these are the kinds of characteristics on which the new élite of industrial scientists establish their status. Moreover, recognising the ambiguity of some of the facts on which the claims are based, and recognising too the history of low status among engineers, helps explain the urgency which was documented in that chapter of the assertion of status and the need for counterposition against another past.

Status and social power

However, as was argued in the discussion of the historical situation, earlier in this chapter, the establishment of social standing draws on wider forces in civil society than those in the immediate process of production. That is to say, while it may be argued that class *location* depends strongly on the place occupied within the process of production, class *character* and *status* are the result of much wider social processes. In the context of the discussion in this book, there are two sides to this argument. The first is that serious reservations must be registered about whether the scientists and technologists working in science parks are in fact working on the scientific frontiers in the way supposed in the rhetoric. In other words, there are questions as to the real content of their place in society's division of labour. This issue has been addressed in detail in chapters 2 and 3. The other side to the argument, however, is that, whatever their place in the division of labour, the status of this group of scientists and technicians is considerably bolstered in the UK by other social processes. These other processes, and characteristics, are complex and mutually interacting, but at this point it is important to highlight just four of them.

First, there is the cumulatively causative fact that these already high-status jobs are overwhelmingly done by white men. Given the society in which we live, this characteristic only further increases their status. In 1985, the EITB (1986) reported, the percentage of women in their survey of those employed as professional engineers, scientists and technologists was precisely 4.5. The IVL survey, mentioned earlier in this chapter, reported sexual discrimination as an important factor turning female graduates away from engineering: '50 per cent said the engineering career structure worked against women succeeding' (Cornelius 1989). Our own data on science parks, reported in chapters 2 and 4, showed a similarly dismal situation there. Numbers generally are increasing, in part possibly as a result of schemes such as the EITB's 'Insight' and the WISE programme, but by 1990 a report on the professions in general showed engineering still to be the poorest performer, with a mere 5 per cent of women amongst its ranks. The situation for black people is less easily discernible from official statistics, but all the evidence points to their absence also.

There are numerous aspects to the overwhelming dominance by men of science and technology. They range from the importance of such dominance in the construction of Western notions of masculinity, and indeed in Western notions of science, aspects which have been explored by a number of scholars, among them Cockburn (1985) and Harding (1986), to the very male culture of engineers, the boys in the laboratory, which for instance Hacker (1981), Noble (1989) and Carter and Kirkup (1990) have addressed. And we have seen in chapter 4 how the design of jobs on the one hand and who does them on the other can become mutually reinforcing and potentially discriminatory.

Indeed, the force of the argument is that the empirical dominance by men of these occupations is not an additional characteristic resulting from the wider social context but a characteristic with something approaching the force of necessity, being inherent in the current social construction of science and technology themselves. None the less, there are some ambiguities in the present situation. Certainly it is the case that, if the new high-status jobs in science and technology are only rarely held by women, the situation was no better in the past. Indeed, it may well have been worse. While the decline of the apprenticeship system and the rise in importance of academic credentialism has reduced the chances for (some) working-class lads to make it up from the shop floor, it may have been – though only marginally – a beneficial change for women and black people. The forms of social closure operated around apprenticeship schemes may have excluded them even more effectively than does the current situation. Thus, in the EITB survey, of the 3,821 women professional engineers in 1985 more than half were working in just two sectors – precisely those of electronics and office and data-processing equipment; in other words, high-tech. The further point is, however, as our data in chapter 4 showed, that even here they are rarely – very rarely – in positions of power and control.

A second way in which the status of scientists and technologists has been increased has in fact applied only to some of them. Indeed, it has served, apparently quite deliberately, to divide one group within the profession from another, to reinforce the new hierarchy. This has been the promotion, during the monetarist years, of a particular form of economic policy and ideological offensive. 'Sunrise' industries have been clearly distinguished from 'sunset', the former applauded (and subsidised) the latter derided, accused of being subsidised, and left to market forces. The public sector has been persistently denigrated and the previous standing of higher education attacked; scorn has been poured, in precisely the older spirit of the 'practical man' deriding theory, on university 'boffins' cut off from the real world (where, presumably, accountancy now rules). And there has been a determined policy of income redistribution, from public sector to private and from poor to rich. Academics and scientists in the public sector had their own means, as have all 'professionals', of bolstering their status in society. What has happened in the last decade is that the bases of this kind of status have been attacked. The hyperbole, the material rewards, and indeed the fond hope of a solution to society's ills and the economy's problems, have shifted to a new focus. All the claiming of status through symbolic counterposition documented in chapter 4 reflected precisely this new set of values, and in many ways science parks capture the essence of it all.

Third, the social power of scientists and technologists has been bolstered quite simply by their very scarcity. There is now a major debate about to what extent and why the United Kingdom has such an apparently poor record in the production of scientists, technologists and engineers. Endless references could be cited from a discussion widely covered not only by the specialist press but also by the general media. Some reports have already been cited earlier in this chapter. Indeed, the debate has become an issue of party politics.

Thus the EITB (1986) commented on how the area of shortage had switched since the late 1970s, when most skill shortages occurred amongst traditional craft occupations and draughtsmen [*sic*], towards shortages in electronic engineering and computer science. In 1987 the National Computing Centre (Virgo 1987) confirmed the particular seriousness of the IT shortage in the United Kingdom. It was reported that half of UK employers in the IT field were experiencing skill shortages in 1985 (Connor and Pearson 1986: 60). Virgo (1987) estimated a shortage of 22,000 IT staff in 1986, compared with a total of 330,000 IT staff in post. 'More than one in five of IT users and one in four of software suppliers are crippled or have their survival threatened by shortages of IT skilled staff ... One in ten of IT users and almost one in eight of software and service suppliers is crippled or has its survival threatened by the poaching of skilled and experienced staff' (Virgo 1987: 8).

Also in 1987, Silberston's major lecture 'Is there a shortage of engineers?', after spelling out and analysing four different types of potential shortage, concluded: 'when I started to write this paper, I was not sure what my answer would be to the question posed in its title. I am now clear that, at the present time, there are shortages of some types of engineer and surpluses of others. In the long-term, however, especially when one defines "shortage" sufficiently broadly, there is no doubt that my answer is [a] definite "yes"' (Silberston 1987: 21). In February 1988, the Institute of Physics reported that responses to recent questionnaires to industry 'make it clear that physics graduates are in very short supply: the demand for their skills is high, steady and unlikely in prevailing circumstances to come anywhere near being met' (Davies 1988: 64). Skills most often mentioned were IT skills, scientific and technological management skills, biotechnology skills, advanced manufacturing technology skills that integrate mechanical and electronic, and new-materials expertise. Also in that year, an NIESR study (Steedman 1988) concluded that 'In mechanical and in electrical engineering work, France today trains between two-and-a-half and three times as many qualified craftsmen [*sic*] and technicians per head of the workforce as Britain to standards which are as high – and often higher – than equivalent qualifications here' (cited in Leadbeater 1988). The Confederation of British Industry backed a report from the Information Technology Skills Agency on the 'information technology skills problem' (Dodsworth 1988; ITSA 1988). And two surveys of employers and employees in biotechnology companies, produced by the Association for the Advancement of British Biotechnology, warned of an impending shortage of staff (*Times Higher Education Supplement*, 18 March 1988). In 1989 the National Economic Development Council warned that shortages of staff with information-technology skills were likely to continue well into the 1990s (NEDO 1989; Leadbeater 1989).

As has already been seen earlier in this chapter, these shortages and recruitment difficulties occur throughout the system of production of scientists and engineers, from school level onwards. They are, moreover, in the final stages worsened by the leakage of trained people to non-scientific fields, such as the City, and to the 'defence' sector. The effects are clearly damaging in a number of ways. The EITB report cited Tarsch (1985: 19) that the short-term effects included companies doing less research and development (27 per cent of respondents cited this effect where the shortage was of new graduates, 36 per cent where the shortage was of experienced graduates) and the delayed introduction of new techniques/products (the corresponding percentages here were 29 and 42) (EITB 1986: 15). And the increasing level of academic credentials demanded only makes any solution to these shortages a longer and slower process.

But another result of this situation of scarcity is the strengthening of the position of these groups within the labour market. A study by Connor and Pearson showed that among the most frequent responses to shortage were salary increases and increases in the intake and sponsorship of graduates (Connor and Pearson 1986).

Moreover this overall national situation of shortage has a particular and somewhat paradoxical geography. The most serious shortages of this kind of staff are in the south and east of England. London and the Thames Valley emerged particularly strongly in the research by the National Computing Centre in relation to IT employees (Virgo 1987: 9). Shortages there were acute, in contrast to other parts of the UK, such as the North East and most of Scotland, where they reported little or no problem. Northern Ireland actually had 'a modest surplus' (i.e. some unemployment). There were other variations both within and between regions. The North West had more marked problems than the North East, Edinburgh than Glasgow, and the southern parts of the Birmingham conurbation, including Solihull, more than the rest of the West Midlands (Virgo 1987: 25). The same pattern emerged in a survey of chemists, for whom demand is less than for many other types of scientist and engineer. The survey found graduate unemployment was far higher in the north and in Northern Ireland, although the overall dominance of the South East was witnessed by the fact that it had the highest number of unemployed chemists, the highest number of employed chemists, and the highest number of job vacancies for chemists (Simpson and Smith 1986). More generally, an analysis of vacancies advertised in the *New Scientist* showed that, of 8,429 in 1985–6, 53 per cent were in the South East. The highest concentrations of jobs occurred in East Anglia and the South East, with Cambridgeshire, Oxfordshire and Berkshire having particularly high concentrations. 'However, for the rest of England, particularly Yorkshire and Humberside, and the West Midlands, the vacancies for scientists per 100,000 population were very low, reflecting the genuine problems of these areas and the poor opportunities for scientists and engineers in these areas.' Even here, however, the situation was far better than in Northern Ireland and Wales (Simpson and Smith 1986: 52). Incomes Data Services produced a special report in 1988 on engineering skill shortages in the South East, and although pointing out that such shortages are not confined to that region argued that they were probably more acute there than in other parts of the country. Again, too, there were variations within the region, with the shortages being greatest in areas to the south and west of London, and particularly in Crawley, Basingstoke and the area around Heathrow (IDS 1988: 6).

Our own survey of establishments on science parks produced a similar result. The greatest shortages were in the south and east (including Cambridge), and also in some of the big industrial cities – there was a high level of shortage reported in Manchester, Leeds and Aston. And one establishment on Cambridge had been recruiting in Dublin when we spoke to them. But the parks with a zero or very low response to questions of shortage – Bradford, Loughborough, Swansea, Liverpool, Glasgow, Bolton and Durham – were all outside the south and east.

This distribution of shortages among scientists, technologists and engineers is fascinating and paradoxical. First of all, the geography of shortage is a mirror of the geography of employment. Even down to a fairly precise sub-regional level the shortages are concentrated where the jobs are concentrated, and vice versa. Moreover, they are concentrated where the highest proportion of higher-level jobs are. This unequal geography is not simply related to the production of scientists, technologists and engineers, however. Some of the biggest teaching departments are precisely in the older northern universities which used to be the intellectual focuses of major industrial regions. Today they are no longer so. At a broad geographical scale, the concentration of jobs and of growth in jobs is overwhelmingly into the south and east of the country. A second point, therefore, is that while in the nineteenth century and the first half of the twentieth the geography of the production of scientists and engineers seems fairly closely to have matched the geography of their employment (at the level of craft workers and apprenticeships the social-network basis of recruitment would to some extent have ensured that this was so) this is no longer the case. The regional geography of the production of scientific and engineering credentials is far wider than the geography of their use in scientific and engineering occupations, and this is increasingly the case the higher up the scientific ladder one goes.

Moreover, the geography of these occupations in the United Kingdom is highly concentrated. A recent survey showed that the UK had the most geographically uneven distribution of scientists in the developed world, with the one exception of the Soviet Union (Davies 1989). There are numerous reasons for this concentration in the south and east of the country, and we shall be discussing them below and in subsequent chapters. But there is one which deserves mention at this point. This is that, in survey after survey, companies argue that one reason (among others) for location in this region is the availability of highly skilled scientific labour. That is, labour is seen to be available in precisely the areas where the most acute shortages are found. This then is the fourth point. But what it indicates is not necessarily irrationality on the part of companies, but clearer thought about the meaning of 'availability'. Clearly in this context availability refers less to the balance between supply and demand than to the overall size (and maybe quality) of the labour market. But what this interpretation of availability implies, in its turn, is a further cumulative process of concentration as the larger labour markets prove the most locationally attractive.

Now, of course, this concentration into the south and east is part of a much wider North–South divide within the United Kingdom. And it is at this level (point five) that the irrationality lies. Thus the IDS survey mentioned earlier reported many firms in the South East which were experiencing shortages of engineers, and found that the North–South divide itself prevented additional recruitment from outside the region.

> Several companies we spoke to had attempted to recruit labour from outside the South-East, but with limited success ... One company ran a recruitment campaign in the Wirral after BNFL made a number of trainees redundant, but potential employees could not bridge the housing gap and, of those that moved, several went on to higher-paying firms in the area. Similar problems faced CF Taylor (Metal-workers), who manufacture aircraft galleys and aircraft structures, when they tried to recruit redundant Westland Helicopter employees from the Yeovil area, but found that people couldn't afford housing in Wokingham. Recruits from the North and Scotland have tried living in temporary accommodation; but the disruption to family life and poor prospects for permanent housing meant that most returned home. GBE Legg has found that, for engineering and computer staff, 'relocation problems (North–South) preclude national advertising on the grounds of cost effectiveness'.
>
> (IDS 1988: 5–6)

Clearly there is some anomaly here. If people cannot afford to move into the south east, how does concentration happen in the first place? There are a number of answers. One concerns change over time – that North–South disparities in house prices vary cyclically and also that there has through the later 1970s and the 1980s been a longer-term increase in the disparity between the two parts of the country, thus choking off any further influx into the south in later years. There is absolutely no evidence, however, that this heralds an equilibrating counter-attraction of the north. For one thing, there is a strong argument that if companies in high-technology sectors find the pressures of southern congestion too great they will move to European locations rather than to the north of the UK. Moreover, the evidence, for instance of the IDS report, was of companies coping with shortage in the short term – for instance, by more overtime – and arguing for something to be done in the longer term about housing in the south east, rather than of considering moving out of the region (although a few have considered this). Another solution to the conundrum of the coexistence of concentration and difficulties of in-migration is that there are differences between occupational groups. In particular it seems clear that it is the higher-earning groups which can most easily cross the North–South divide and indeed the highest-level jobs which are most concentrated into the south and east. The implication of this, of course, is that North–South disparities, this time in occupational structure, are yet further reinforced.

Moreover, finally, and to come full circle, the continuing higher level of shortages in the south and east of the country reinforces the bargaining power of those who have jobs there in comparison with people with similar qualifications in (or looking for) similar occupations in the north and west of the country. There is abundant evidence of higher salaries, often individually negotiated (e.g. IDS 1988) and of higher rates of turnover among these staff (Virgo 1987; IDS 1988) in the south and east. The National Computing Centre's report of 1987 found that in London and the Thames Valley there were turnover rates of over 40 per cent per year in some categories of IT staff (Virgo 1987: 9). Savage *et al.* (1988: 461) reported turnover in Berkshire among workers with computing skills at about 50 per cent a year. It is those in the higher-level occupations and in the south and east of the country, therefore, who are benefiting most from current national shortages. Moreover, the indications are that the north and the south of the country are becoming increasingly closed-off from each other in labour-market terms. There is evidence that graduates move to large labour markets at an early stage in their career, and then stay there, able to change jobs without changing region (Savage *et al.* 1988). If it is a widespread phenomenon, this can only reinforce the cumulative clustering of higher-level jobs in the already larger labour markets of the south.

We have seen, then, some reasons for the concentration of high-level jobs in science and engineering in the south and east of the country. But there are other reasons, too. Thus it has in part been made possible by the increasing separation of conception from execution, of mental from manual, which has been documented earlier in this chapter. It is a long-term stretching-out of the technical division of labour within production which has enabled (not necessitated) spatial separation also. A more positive inducement to southern concentration has come from the state itself. It is now widely demonstrated and agreed both that a disproportionate share of government research activity and of the recipients of governmental scientific contracts (e.g. for defence-related work) is located in the south and east and that this has had an important propulsive effect on the growth in that region of high-technology sectors, including perhaps most particularly but by no means exclusively electronics. Most recently, and very pertinently in the context of a discussion of science parks, a parallel argument has been put forward about funding – and changes in funding – the higher education system:

> Last, but hopefully not least, are the universities and polytechnics. The Robbins era of expansion in the Sixties succeeded in spreading academic talent remarkably fairly around the country. But the swingeing cuts suffered in this sector over the last few years have fallen far from evenly. Universities in Scotland, Wales and the North of England have by and large borne the brunt of the cuts. In the latest round, for example, five universities suffered cuts in actual cash terms: only one of these was in England.
>
> (Davies 1989)

Davies (a professor of theoretical physics) predicts a future when non-southern universities could become teaching-only institutions, with research concentrated in the south. 'Northern universities thus risk going the way of northern industry, relegated to a subsidiary, metal-bashing role.' This crucial role of state spending and state policy in the North–South science divide will be examined further in chapters 6 and 7.

But there is a third element in the explanation of the concentration of these jobs in the south and east, an element equally commented upon and agreed. This concerns the preferences of the employees themselves. There are good labour-market reasons, as we have seen, why such employees might prefer a location in the south. But there are also social reasons. Wider and more integrated social networks, access to a greater range of their preferred cultural resources, greater access to sources of power and decision-making, an overall environment associated with growth rather than decline and, last but by no means least, distance precisely from production, or certainly production of the factory-based, dubbed 'sunset' kind, and its associated workers – all these things attract these groups to the south rather than the north. So for reasons of social status and security and of career progression the south wins again. The class power which these groups already have, in other words, enables them to a greater degree than in the case of other workers to make their own choices about where to work. And that choice, as we have seen, if it is for the south can further reinforce their social standing and negotiating strength within the labour market.

This brings us, then, to the heart of the fourth element in accounting for the status attached to these occupational groups. This element revolves around space and place, around geography. And here, we would argue, science parks epitomise the processes at work, and their effects. Indeed, spatial form and spatial content are integral to the definition of science parks.

In the early chapters of this book we argued that the popular conceptualisation of science parks is a rough-and-ready one which gives little purchase on the real causal processes involved. We argued, therefore, for a more thoughtful approach to their conceptualisation and proposed that there were three elements to such a reconceptualisation. The first of these three elements was the linear model of science and innovation introduced in chapter 3, on an assumption of which science parks are based.

The second necessary defining characteristic of the archetypal science park is that it has a particular spatial form and content. There are three distinct elements to this spatial aspect of the definition. The first two of them concern the spatial organisation of the linear model of scientific production and innovation, and its associated division of labour (see figure 3.1). They imply, in other words, a particular spatial structure of production (Massey 1984) for this model. The first element is that there should be spatial contiguity/closeness between those elements of the linear structure located on the science park and academe. In other words, in relation to figure 3.1, if the science park is, roughly, boxes 2 and 3 (applied research and experimental development), then these activities and their associated personnel within the overall division of labour should be physically located close to box 1 (basic research), which is in academe. The second element is that any major physical production activities, and the associated employees, should definitely not be on the park but should be located 'elsewhere'. In the terms of figure 3.1, in other words, there should be geographical distance between box 4 (full production where this is of a physical nature) and the activities and personnel further upstream. In the very definition of a science park, in other words, there are definite requirements both of contiguity and of separation in the spatial organisation of the linear model.

The third element in the spatial characterisation of the archetypal science park concerns what might be called 'spatial content'. This is that the symbolic spatial content of a science park should exude exclusivity and status, as part of its design.

Two points should be made. First, it will be noted that all three elements of this spatial aspect of the conceptualisation of a science park (contiguity, separation, exclusivity/status) are once again necessarily relational. That is, they are not autonomously descriptive characteristics: they necessarily have implications for the geography of others not on science parks. Second, as in the case of the linear model, by setting this aspect of the conceptualisation of a science park in the context of our wider empirical studies a number of further implications can be drawn out.

First of all, a caveat: as in the case of the linear model it is again true that this aspect of the archetypal definition is not always and everywhere precisely realised in practice. As we have seen in chapter 2, there is some physical production on science parks, although very little, and there is considerable activity which is not serious R&D (there are marketing outlets, listening posts, warehousing, etc.). There is also production of some products such as software.

Second, and again as has been demonstrated in chapter 2, the contiguity element of a science park is not producing its hypothesised (and sometimes actually claimed) effects. These are not primarily the sites of academic start-ups, or of the development of products derived from research in the associated HEI, nor is there a great amount of research linkage between park and HEI.

But, third, if these popularly postulated effects of the spatial form and content of science parks do not seem fully to be being realised, each of the three spatial elements in our reconceptualisation *does* seem to imply social processes, but social processes *other than* those claimed in the promotional literature. The element of contiguity to academe is most certainly reinforcing the élitism of industrial R&D. While this may seem ironic at a time when higher education itself in the UK certainly feels under the threat of declining status, the evidence from both textual analysis and from our interviews is indisputable. The importance of self-definition as 'university-related', or 'lab-like' (and many other such phrases) was constantly evident. In one of our interviews, at the science-park establishment of a multi-site multinational, the interviewee pointed out that while elsewhere in the company QSEs were called 'engineers', on the science park they were called 'scientists'. Moreover, the exceptions made for allowing 'production' on to science parks, most especially software production, indicate the essentially social, in the most general sense, character of the exclusion. Other examples have already been given in chapter 4. This factor also emerged quite precisely from our survey of reasons for the choice of site by establishments on science parks. 'Prestige of being linked to the university' was the third most often mentioned factor out of a list of twenty-three factors mentioned.

Equally evident was the importance of the element of separation – separation from physical production. Inevitably this emerged most strongly in the unstructured in-depth interviews. Indeed, as has already been argued in chapter 4, the frequency with which science parks were defined as prestigious precisely through counterposition with an assumed inferior-status and spatially distanced 'other' (usually manufacturing industry, big firms and production workers) turned this aspect of them into a sub-theme of our investigation.

Finally, all this status definition was further reinforced by the symbolic spatial content of science parks' design, and the way in which their design was written about in the promotional literature. In virtually every case, whether it be in semi-rural areas or industrial districts or inner cities, the defining characteristic of the physical design of the parks and the way in which they were promoted, was that of high status, a prestige environment in relation to the surrounding area. In our questionnaire survey on location the factor 'prestige and overall image of the site' was far and away the most often mentioned, being noted more than one and a half as many times as the second factor. Moreover in the 1990 follow-up survey, when science-park establishments were asked to name the top three factors, relating to the science park, which had contributed to their success over the last four years, the factor which ranked ahead of all others was – 'prestige address' (Newland, at 1990 UKSPA Conference). And finally, in a 1990 survey of forty-six companies on Aston Science Park, 'image of science park/prestige address' was the most often cited reason for location, while a question on perception of the benefits of a science-park location elicited an 83 per cent response on 'high' benefits for image (with 15 per cent 'medium' and 2 per cent 'low') – the figures for 'university links' were 28 per cent, 22 per cent and 50 per cent (Birmingham City Council 1990).

These implications of the spatial form and content of the archetypal science park all therefore reinforce the tendencies towards social inequality already apparent from preceding chapters in the discussion of the linear model of science and the organisation of work. The attraction of science parks as prestige locations for some establishments is not only to enable them to attract 'the right staff' (as it is so frequently put), although it is often that; it is also to establish a position of status in relation to other companies. Whatever the reasons, however, the effect is to confer some aspects of that status on the people who work there.

But there are also implications for technological development and innovation. Most particularly, the spatial separation of R&D from direct physical production is increasingly argued to have potentially negative effects, as indeed is implied by some of the reservations currently being expressed about the linear model (the greater importance of learning-by-doing, etc.) (see chapter 3). This is of course even more the case where, as is typical, the activity on the science park is not near the 'basic research' end of the spectrum (chapters 2 and 3). There is also a more general problem. In the UK, as we have seen, one of the important background arguments for the establishment of science parks was that 'the British' are (supposedly) good at research break-throughs and fundamental inventions but bad at commercialising them.

This argument is related to that disdain for physical production which has been such an important component both over the long term in the maintenance of social status and more recently also in the determination of the unequal geography of high-tech. Science parks were supposed to aid in bridging the gap between the academy and the commercial world. Their impact, however, might rather be to tie one group of scientist-technicians within industry to the academy (although in spatial and social terms, rather than through scientific and technological interchange) while reproducing within industry itself the correlation between status and distance from physical production. If this were to happen at the same time that basic research (the source of the previously uncommercialised break-throughs) was itself undermined through lack of funds and political attack, the irony would indeed be deep.

NOTE

1 Wright himself argues this in a more recent paper (1988). It also re-poses in a rather different way the issue of trajectories discussed earlier.

6

SCIENCE PARKS AND LOCAL ECONOMIES

Few journalists have been moved to write, as they have in the case of Cambridge, of the ride to the science park at Aston. The flights of fancy which we reported at the beginning of chapter 4 are not repeated here.

So we decided to do the journey ourselves. You can get to Aston easily from almost anywhere in the UK (unlike Cambridge, which is virtually only reachable via London). Coming by rail you arrive right at the centre of town (the station in Cambridge is on the outskirts). It is a town centre moreover which was rebuilt, completely refashioned, in the 1960s and is due to be rebuilt again. None of the conservationism of Cambridge here. Instead of preserving the past, this place is pulling it down and replacing it. From the station you thread under motorways through subways with their familiar sour smells and draughts, decorated by graffiti, to arrive in only a few minutes at 'The Triangle' of the science park.

Or you can arrive by motorway. Once again in contrast to Cambridge, the motorways here, in a city which for a few decades prospered from building cars, plunge right into town. You curl along flyovers, slice along highways cut through an older urban fabric, and a few minutes from the end of the motorway you are at the science park.

The science park, and the university, announce their presence by the Aston Triangle. This is the site of the science park. And everything about it, from the architecture to the security checks, asserts its difference from the West Midlands conurbation which spreads out around it.

Although the park thus struggles to state its distinction (which, as we shall see, has its own ironies, since the aim – and practice –

is to link into the local economy) the overall setting could not be more different from the normal rhetorical imagery of high-tech locations. For the imagery of science parks has been dominated by Cambridge.

Yet, as we shall argue in this chapter, not only is Cambridge a highly specific (probably one-off) case but, and more positively, 'science parks' can lock into their local economies in a whole variety of contrasting ways. In that sense, the dominance of Cambridge images has been a hindrance rather than a help. For science parks in different areas are responses to, and lock in to, distinct local circumstances. Yet, in all cases, 'image' is essential.

In this chapter we explore in detail two cases (Cambridge and Aston) to point up some of the potential dimensions of these contrasts, and conclude by assessing more widely what such local specificity implies for the potential of science parks as part of a local policy package.

The approach we adopt is based in a particular way of theorising local specificity. In particular, it conceptualises local economic specificity in terms of the superimposition over time of a series of rounds of investment, with each succeeding round of investment being in part a response to, and interaction with, the particular characteristics which are the inheritance of the combination of previous rounds. In that context, the investment in the science park may be considered as being (part of) a new round, though the nature of it has to be established. One question, then, concerns the way in which the science-park investment relates to the historical inheritance. Each round of investment, however, is wider than the local area and within each round the area will play a particular role. Previous rounds will therefore have established Birmingham and Cambridge in particular positions within the wider division of labour. A further avenue for exploration is therefore how this role may be changed by the new investment represented by the science park. Finally, there is the question of the detailed economic and social links between the science-park establishments and their local area more generally.

These questions are important for both theoretical and policy reasons. The theoretical questions revolve around the issue of uneven development. If we are indeed, as the terminology of new technology implies, at a point of change in terms of industrial development, then how do the geographies of these different eras interact? Science parks provide an ideal focus for studying this question. It is also precisely because there is uneven development that the policy questions arise. Different phases of industrial development have different (unequal) geographies. One of the questions which underlie this chapter is how best local economic strategies can relate the new rounds of investment to their existing economic bases.

THE ECONOMIC HISTORY

Aston

The history of capitalist industrialisation and economic development in the UK over the last two centuries has been marked by shifts in the regional focus of production. Lancashire led the earliest wave, with the beginnings of factory production in textiles; the coalfields of South Wales and the north east dominated the era of coal mining and steel and railways. In similar vein, the outer areas of the south and east of England are the focus of today's growth, including that whole bundle of activities broadly known as today's 'new technology'. As Marshall argues, 'these evolving patterns of uneven regional development are not simply reflections or outcomes of the long waves in the national economy. They *are* the process of national economic change and development' (1987: 228).

A century ago it was the West Midlands, including Birmingham, which was establishing its career as the engine-room of national growth. It was an expansion much aided by public-sector intervention. As the Economic Strategy for Birmingham (1985–6) is quick to note:

> The City Council played a major part in providing the infrastructure for this period of growth. The municipal enterprises started by Joseph Chamberlain were the foundation on which the City was able to attract and keep the major growth industry of the first half of the twentieth century – the motor industry.
>
> (Shaylor 1985: para. 2.1)

Imperialist profits were ploughed back into the municipal administration, public works, social infrastructure and architectural projects of Birmingham's 'civic gospel' and potential working-class discontent was thereby lessened (Marshall 1987: 178, citing Briggs 1968).

For long a constellation of small workshops in a myriad of trades, by the early part of the twentieth century the region's economy was focused on a small number of sectors (electrical engineering, machine-tools, cars) and, in time, a small number of larger companies surrounded by flotillas of tiny firms many of which serviced them. This crystallisation-out of the twentieth-century economic structure of the Birmingham economy also brought with it a new form of social organisation of production – Fordism. It was a manufacturing area which from being the regional base of Britain's reputation as the workshop of the world evolved into the home-base of a group of powerful multinationals. In the second half of the 1970s about 100,000 workers, that is, over 40 per cent of all manufacturing workers in Birmingham, were employed by a mere ten companies, and the thirty manufacturing establishments at each of which more than 1,000 people worked accounted for 39 per cent of the city's total manufacturing employment (Birmingham City Council 1982). Moreover, this was no global outpost of multinationals based elsewhere. Of the top ten private-sector companies in the West Midlands, ranked in terms of the size of their world sales, no fewer than six (GKN, Lucas, TI, IMI, Glynwed and BSR) still in 1982 had their headquarters in the region (the other four had their headquarters in London) (Gaffikin and Nickson 1984: table 4.4). Similar calculations on 1977 data show that, of the ten largest private-sector companies in terms of their employment in the region, five had their headquarters there. Moreover many of these firms were 'traditional' 'West Midlands firms with roots stretching back to nineteenth-century family businesses' (Marshall 1987: 215). Out of all these companies only one, Talbot, had its headquarters outside the UK. Indeed, quite the contrary: the West Midlands base of these companies is the centre of vast networks of overseas subsidiary and associated companies (Gaffikin and Nickson 1984: appendix B).

The West Midlands, then, was a main centre of production for a whole century of national economic growth. It was moreover, indeed still is, a centre of the ownership of production in those previous rounds of development. It was a region in which was based an important part of the UK economy's role in the international capitalist economy, and a region from which many of the tentacles of imperialist and subsequent control spread out around the globe.

It is with the inheritance of this success that the people and the local councils of the region are currently trying to cope. In the 1970s, of all the regions in the UK the West Midlands was one of those in which manufacturing most dominated the employment structure. As such, the economy of the West Midlands and specifically of the conurbation did not so much reflect as in large measure embody the increasingly rapid deindustrialisation of the UK in the years after the late 1970s.[1] It is estimated that between 1979 and 1983 the West Midlands economy lost one-third of its manufacturing jobs (West Midlands County Council Economic Development Committee 1984, cited in Marshall 1987: 215). Moreover a significant part of this loss was attributable to the international restructuring of the major multinationals in the region. The British share of employment in the ten major companies in terms of global sales fell from 75 per cent in 1978 to 67 per cent in 1982 (Gaffikin and Nickson 1984: table 4.6).

The West Midlands, then, was the regional focus of previous rounds of national economic growth. It was a centre of accumulation. Today, the spreading international basis of its home-grown company empires is clear witness that a region's 'having its own headquarters' is no guarantee of retaining jobs. This is a region which is losing its production as the recent decades of international restructuring both hasten and mark geographically the long historical process of the separation of ownership and control from production. As the city council's review put it 'Birmingham's poor investment performance in part reflects its relatively high dependence upon large multinationals, for whom overseas investment and growth is an option not available to small and medium-sized companies. This process of overseas investment has not been offset by inward investment into Birmingham from overseas-based MNCs' (Birmingham City Council 1986: para 2.4).

The result was a massive rise in unemployment, with ten wards in the city having rates of over 30 per cent in 1984 (WMRSR 1985: table 1.13) and, for those who still had jobs, regional average earnings for full-time adult manual workers falling from the highest in the country to eighth (for men) and from second to seventh (for women) between 1972 and 1982. As late as 1990 unemployment in inner Birmingham was still twice that of the West Midlands County, which in turn was one and a half times the level of the UK as a whole (Birmingham City Council, Area

Unemployment Summary, October 1990).

When Aston Science Park was established in 1983, the economy of the area was still biased (in comparison with the national average) towards manufacturing, and within that towards a heavy concentration in a few sectors, and towards employment in a few large companies. Unemployment was well above the national average. And the occupational structure of the region/city reflected its history, with a high proportion of manual workers and low numbers of professional and scientific workers. Trade unionism was well established. There was little in the way of greenfield locations or areas of 'parkland landscape'. Fifty per cent of the industrial building stock was over fifty years old (BCC 1986: para. 2.2).

The science park was backed by Birmingham City Council, Aston University and Lloyds Bank. Birmingham City Council, interestingly in the light of the city's industrial history, adopted an approach to strategy founded on export-base theory. 'The future of the City's economy depends on the health of its basic industries (i.e. those with a high export content – mostly engaged in manufacturing). These are the industries which generate the City's wealth and upon which many other jobs depend, particularly those in the service sector' (Shaylor 1985: 1). What was necessary was to revitalise the existing export base, much of which through lack of investment was still languishing in a previous technological age, but perhaps more importantly to encourage the growth of new regional exports (which, it was recognised, might well include some services). Among the main obstacles to the achievement of these aims the council recognised as particularly important the lack of suitable land and buildings and the problem of 'image'.[2] Both were clear inheritances of the area's previous 'success'. Within this context, the role of the science park from the point of view of the council was to be part of the programme of restructuring existing basic industry and developing new basic (i.e. exporting) industries through the introduction into the economy of new-technology industries (BCC 1985; Shaylor 1985). Direct job creation was certainly a consideration, but it was explicitly seen as a more long-term aim (BCC 1985). The same document also argues, first, that while links with local industry and academe may be important they should not be seen as exclusionary criteria, and second that social criteria should also guide decisions. Again, in the 1986 Economic Strategy Document, the science park is seen as important in both the land and buildings programme and the technology programme. It is to be 'the leading technology location in the Region' (1986: para. 5.5).

In both these programmes, however, the problem of the distribution of benefits is recognised. Thus: 'The Council is aware, however, that land and development initiatives on their own do not necessarily provide significant benefits for the most economically disadvantaged sections of the community. Greater emphasis will therefore be placed upon linking land and buildings programmes into wider economic strategy goals concerning, for example, unemployment and community based initiatives' (para. 4.17). Here, then, is an attempt to confront the problems of social inequality which as we have seen are inherent in the very definition of the archetypal science park. But how was this aim to be reconciled with the characteristics of high technology? For alongside the socially egalitarian intentions 'The Science Park remains the innovative cornerstone of the City's technology potential. Its importance to the image of the City in promoting Birmingham as a technology user and producer cannot be over-emphasised' (para. 8.8).

'Image' was probably also important in the reasoning of the academic partner in the establishment of the science park. Aston University is not an élite establishment. It began life as Birmingham College of Commerce, which became a College of Advanced Technology. Subsequently created a university, it none the less tended to remain rather in the shadow of Birmingham University, an old-established and very large red-brick institution, located in the relatively leafy suburbs of Edgbaston. Aston, in contrast, is in the heart of the declining inner city. It suffered the second largest funding cut in the 1981 round of attacks on higher education. As well as the park and its companies gaining kudos from proximity to the university (something we shall explore later), the relationship also works the other way. In the first half of the 1980s Aston University, under its Vice Chancellor Sir Frederick Crawford, went through a major rethink. Crawford, new to the job and just returned from Stanford, set about a major switch in the nature and image of the university, of which the development of the science park was part. Aston University had no land or endowments to put into the project (the contrast with Cambridge could not be more stark) so in this 'city of civil enterprise' (interview: science-park management) it turned to the local council. In the university itself, entry grades were raised, ten departments were closed or merged, humanities and social sciences were dispensed with, the student population was reduced (including halting all non-A-level entrants), courses were dropped and staff reduced (Rogers 1985). 'Emphasis is strongly on programmes that better meet future needs of industry and commerce' (Rogers 1985). Integral to all this was the physical development of the campus and the reworking of an image. International design consultants specialising in work on corporate identities were brought in (Rogers 1985): buildings were demolished, others renovated, a main road through the centre of campus was to be diverted, there was a new logo. Advertisements for posts at the university mentioned the science park in their blurb. Said a park liaison officer in interview, 'The park is the vision of the university in the future'.

But the provision of land and buildings on their own was judged by all the partners to the science-park venture, and quite correctly, to be inadequate as an industrial strategy to overcome the problems faced in the city. Aston Science Park is probably one of the most 'pro-active' in the country although in recent years the pro-active role of BTL seems to have diminished, as does the active promotion of contact between park residents. But in the early years there was a range of services available, an active management, a Lloyds business consultant, and perhaps most important a source of investment finance – Birmingham Technology Ltd. Even physically preparing the site demanded considerable action and finance, and, like everything else, reflected the economic history of the region in which the science park is set. For the first park buildings are located on an old site of Delta Metals, where previously there had been a non-ferrous rod mill. The change in land use was thus symbolic of the aims of the economic strategy itself. Moreover, it involved a physical change which equally epitomised the social processes under way. The problem was the removal of the massive foundations of the mill. And, as usual, the public sector, in this case in the shape of the council, using Urban Programme funding, picked up the tab for quite literally digging up and removing the remains of the past economy to provide a basis for new rounds of (private) accumulation.

Cambridge

The Cambridge story could hardly be more different from all this. It is also well known and often told and we have already seen something in chapter 4 of its almost mythological status. What we shall concentrate on here are the bare bones of the structural differences from Aston.

In the second half of the nineteenth century, when Birmingham and its region were gearing up to take over the lead in national patterns of industrial development, Cambridge, so far as accumulation was concerned, was a backwater. It had an internationally famous university and the market functions of its rich agricultural hinterland. Certainly in this period two companies (Pye, and the Cambridge Scientific Instrument Company) were formed and their birth is always treated as seminal in those narratives which seek origins in continuities and early examples in the past. But at least equally significant to what was to happen later was the fact that Cambridge did *not* become a centre of nineteenth-century industrial development.

The city stayed that way through the first half of the twentieth century. The most important development, during this time, in terms of laying down pre-conditions for the future 'phenomenon', was the location in the city of a whole series of specialist research institutes. The Plant Breeding Institute was established in 1912, the National Institute of Agricultural Botany in 1919 and the Institute of Animal Physiology in 1948.

The post-war period, too, looked set to continue the pattern. Most famously, the Holford report, commissioned by the Labour government's Department of Town and Country Planning, financed mainly by the Conservative county council, and accepted by the Conservative government in 1954, argued that Cambridge's 'special character' must be preserved. What this meant was that pressures for 'unsuitable' development should be resisted, and in particular that large-scale manufacturing activities were to be discouraged from establishing anywhere in the county (Segal Quince 1985: 18). Even proposals for large-scale research facilities were turned away, including applications from Tube Investments and IBM. A 'green belt' was drawn closely around the city and Industrial Development Certificates were used in the 1960s to control growth.

The Mott report, moreover, written at the end of the 1960s, linked together the social evolution in those years of the scientific community, discussed in chapter 5, and the politics of geographical location. It started explicitly from the linear model of scientific discovery and innovation which we discussed in chapter 3. Thus it wrote 'It is important to recognise that the steps by which an invention can be translated into an innovation involves [*sic*] a continuum of research, design and development and prototype production through to manufacturing and marketing functions which are highly sensitive and easily disrupted in the early stages'. While the early stages could be in Cambridge, it argued, 'A company can move its production units to other areas particularly when this requires a different form of labour' (*Cambridge University Reporter* 1969: 374). It continued: 'What is required, therefore, is some definition of the stage at which a firm is capable of separating its production units from its research and design laboratories. The best definition the Sub-Committee can give of this stage is where production routine does not require daily discussion and action on the part of laboratory and design staff and where the type of labour employed does not require lengthy and formal training in scientific and technological skills' (p. 374). However, this is a very fine line to draw and 'existing instruments for planning control are too crude to provide a satisfactory solution to this problem' (p. 373), so another solution had to be found: 'The County authorities have stated that these industrial [planning] applications present the real difficulty when endeavouring to establish planning controls which would prevent research-based units from developing into large-scale manufacturing processes of a type which neither the University nor the County believe should be located in Cambridge ... The Sub-Committee believe the most effective form of control is that exercised through leasehold conditions which could be applied through the creation of a "Science Park" sited in a position which allows good access to University Departments' (p. 373). This particular link between the model of science and innovation, its social implications and its geographical form in a science park could not be more clear.

It was not until the approval of the Mott report in 1969 that policy changed. This report took issue with the Holford doctrine and argued that a limited growth of science-based industry and of developments related to technology and research would be beneficial. There were a number of reasons for this decision to relent, if only in highly defined (and highly refined!) ways, to the mounting pressure for expansion. A range of arguments came from the university, long the protagonist of the 'conservation' argument, and in particular from the physics and engineering departments, which had noted what was happening across the Atlantic, at Stanford and MIT. One of these arguments – that while the theoretical standing of the university was widely acknowledged, its contribution in the applied field needed to be stated more clearly (*Cambridge University Reporter* 1969: 372, 376) – provides an interesting contrast with the situation at Aston. But there was also a wider argument in these years of the white heat of technology, that a more mixed environment would be stimulating (Segal Quince 1985: 19), and that relevance to industry was important to the national economy (*Cambridge University Reporter* 1969: 371).

These two shifts, in Cambridge University and in national arguments about technology, were related both socially and institutionally. Sir Neville Mott was head of the physics department at Cambridge (the famous Cavendish Laboratory) and a solid-state physicist. This was the research area which had produced the transistor and the computer 'chip', technical base of the computer and the prime technology of the 'clean' electronics industry. The idea of a new kind of industry, modern and clean, had been integrated into British science and technology policy in the late 1950s and 1960s, particularly by a group of scientists who were scientific advisers to the Labour Party in opposition. Two principal members of the group, unofficially called 'The VIP Club', were ex-Cavendish Laboratory scientists, J. D. Bernal and Patrick Blackett. Their advice, together with that of C. P. Snow and Solly Zuckerman, culminated in Harold Wilson's call at the 1963 Labour Party Conference to 'forge a new socialist Britain in the white heat of the scientific and technological revolution'. Blackett became President of the Royal Society and then, following the election of a Labour government in 1964, Chief Scientific Adviser to the new Ministry of Technology. It was this Ministry that in 1969 set up the CAD Centre, a government-funded facility managed by the UK's top computer company, and which chose to locate it in Cambridge near to the university.

A whole number of things thus come together here. The growing divide between mental and manual within the division of labour, documented in chapter 5, was part of what enabled the spatial separation of full production on the one hand and R&D on the other. The increasing importance of academic credentials reinforced the links between the latter and the universities. This more strung-out division of labour together with technical changes such as the development of the transistor and the chip lent a very different image to the new generation of industry, one which even preservationists of the city might see as not unsuitable for location in Cambridge. But – or moreover – as we have seen both in this chapter and in chapter 5, the net result of all these shifts was also very anti-egalitarian. This was true both in the clearly hierarchical division of labour, which became even more firmly entrenched, and in the social networks, and the locations, in which the new industries and the new policies were developed. It was a classic 1960s Wilsonian technocratic 'expert' solution. Potentially progressive in the intent to educate more scientists, to stimulate public investment and intervene in industry (and we shall see in the next chapter how important such public investment has been to local economic growth), even to introduce industry into the hallowed atmosphere of Cambridge, it also bore the problems of social inequality endemic in technocratic solutions.

The type of new development to be allowed in Cambridge was thus quite tightly specified; it was certainly to be 'science-based'. As Segal Quince (1985) put it 'the process of preparing the document helped clarify the distinction between "smoke-stack" and "science-based" industry and legitimised the role of the latter as an integral element in the future of the University's research and the Cambridge scene more generally' (p. 21). And as the Mott report put it 'A great deal of industry may claim to be "science-based" in some sense and there are examples of companies establishing laboratory facilities more with a view to meeting planning requirements than as an essential element of their industrial activities. On the other hand, there are also examples of research laboratories employing a thousand or more scientists and technicians. Both types of development would, in the Sub-Committee's view, be undesirable for this area' (pp. 373–4). In other words, what was emerging as the new kind of industrial development might not be so out of keeping with the social ethos of the town as defined by those who saw themselves as its leaders and protectors. The impetus for the report, like Mott himself, came again from the university: 'it was almost certainly without precedent in Britain at the time that a university should take the lead so explicitly and forcefully in such planning matters – it is still highly unusual even today' (Segal Quince 1985: 21). The report explicitly suggested the creation of a science park.

But there were other forces, too, pushing for a change in the old 'anti-development' policy. Among them was the pressure exerted through the city council, which came under Labour control in the mid-1960s and in whose area the initial part of the science park was located.[3] The concerns from this direction were very different. Although unemployment was historically low in Cambridge, there were employment problems and particularly there were problems of low pay. By far the largest section of the workforce was employed by the university, with, apart from Pye, the other sources of jobs being mainly shops, hospitals and local services. The dominance of the university in the town was, in other words, part of the cause of the problems experienced by the unemployed (through its anti-industry stance) and low-paid (through both that and its own employment policies), and perceived by the Labour council. And the perpetuation of those problems had been ensured by the historically dominant 'conservationist' power bloc which effectively held down both the city's rates income and its range of job opportunities.

Finally, the low level of unemployment (relatively) was also producing 'recruiting difficulties' for some major local employers, 'which they argued resulted from unwarranted restrictions on provision of housing' (Segal Quince 1985: 19).

Each of the three sets of interests, or representatives of interests, then, wanted expansion. But, what is not pointed out in most accounts, they each wanted a different kind of expansion. The university specifically wanted properly 'science-based' industry, the city council wanted more jobs and higher wages for working-class people, and also a higher rates income, while some local employers wanted housing (i.e. more people rather than more jobs) in the area.

These were times when considerable changes were happening anyway. In the wider context of national patterns of manufacturing employment a process of geographical decentralisation was under way, from metropolitan to less urban and from 'south' to 'north'. While the West Midlands lost out in this process, the region around Cambridge – East Anglia – gained. In parts of the 1970s it was indeed the only standard region to record an absolute increase in manufacturing jobs. Moreover this was not on the whole the decentralisation of major branch plants as it was in the northern (regional policy) regions of the country. Far more, the new investment consisted of smallish firms often moving lock, stock and barrel from the metropolitan area to the south.

The economic and social history of Cambridge, in other words, stands in complete contrast to that of Aston/Birmingham. This was no international hub of previous rounds of industrial development or entrepreneurship. There was no dominant industry, nor even a dominant company. And all this is frequently hypothesised to have had social and economic effects conducive to the growth of the current round of investment. Thus:

> there was no dominant industry or company. So local people set up companies to sell to niche markets. People in Cambridge did not regard themselves as employment fodder.
>
> (Segal, quoted in Eustace 1985: 20)

and again:

> the fact that there has never been heavy industry, or industries in which large plants and large unionized labour forces have been prominent, has helped create a labour market and a general attitude in which flexibility and individualism have never been suppressed. A history of low wages – due to the long dominance of the agricultural and low-level services sectors (the latter partly a result of employment patterns in the University and colleges) reinforced by the early industrial employers – and a generally low penetration of trade unionism have contributed to the effective functioning of the labour market.
>
> (Segal 1985: 566)

On the other hand, but again because of the university, there are:

> numerous interlocking networks of talented, influential and accessible individuals
>
> (Segal 1985: 566)

and:

> Quite apart from the sheer quality of the University is the matter of 'style'. There is an assuredness – born of a mix of a long and distinguished history, substantial resources and a sense of uniqueness.
>
> (Segal Quince 1985: 53)

However much credence one gives to these precise points (and we would certainly agree with Segal's insistence on highlighting 'the importance of history in "conditioning" subsequent economic change'; Segal 1985: 565, and see Massey 1984) what they are indicative of in the Cambridge case is the significance of the pre-condition of social inequality. Examination of figures in the New Earnings Survey shows clearly that the gap in earnings was consistently greater in the Cambridge area than in the West Midlands, especially for men, through the 1970s and 1980s. It clearly recalls the characteristics described in chapter 4. What these accounts of the Cambridge phenomenon point to (usually inadvertently) is the significance of both an élite, 'in crowd' network on the one hand and a low-paid, non-unionised workforce on the other. (The key role of Cambridge University in the construction of both raises from a rather different angle from usual the importance of a local HEI in the growth of high-technology industry.)

But these are among the particular conditions which are conducive to the growth of the new industries of the present era. What the Aston–Cambridge contrast throws most into high relief is the more general phenomenon of shifting patterns of spatial uneven development as the economy moves between historical phases. Old areas are abandoned for new. It could not be clearer than in the contrast between the property sectors in the two places. The built environment is a key element, along with others such as labour, in shifting patterns of uneven development. But it is also stubbornly fixed in space. It can moreover quite rapidly become inappropriate. And it is also – as we shall see very clearly in the next chapter – itself a source of accumulation. In Aston/Birmingham property was a problem. There were acres of empty property but little of it was suitable for new high-technology companies. The provision of new property was seen as one element in a package designed to generate demand, and it was provided mainly by the public sector. In Cambridge the supply of new property has in large part been a response to a demand which already existed, and the supply has been new-build, and largely by the private sector.

The Cambridge Science Park fits squarely into this overall context. It is an initiative of, and funded by, a college of the university (Trinity) rather than the local authority. Trinity's objectives 'give equal standing to the advancement of science through commercial applications and the financial return on their investment' (interview); there is no mention here of local economic development, job creation or wider social objectives. The science park has not produced the Cambridge Phenomenon, rather it is part and parcel of it (Segal 1985: 566): 'The College and Bidwells do not actively canvass for new tenants; demand for the units is market-led, the philosophy being that they are meeting a need not creating it' (interview). Indeed, the park has been estimated to account for only 10–15 per cent of the whole Cambridge high-technology presence. And instead of, as in the Aston case, the public sector having to spend millions of pounds rehabilitating the area from the effects of previous rounds of accumulation, in Cambridge the college negotiated for four years to get planning approval (which is still dependent on a 'use clause' being incorporated into every tenancy; interview) because it is inside the local 'green belt'.[4]

PARK LINKS

Given, then, those very different contexts of economic and social history, how have the two science parks developed in relation to their local areas and the wider international economy? It is valid here to draw some fairly direct comparisons between the firms on the two parks, because, while the previous rounds of investment in their local areas are highly contrasting, and although the two parks are as we have seen each part of a rather different broader policy context, it is none the less true that they are both attempting to establish rather similar types of new investment.[5] Both have targeted investment which is characterised by adjectives such as 'innovative', 'science-based', 'high-tech' and having links with the university. And both have succeeded in promoting such investment, broadly defined; both parks are now quite well established. However, the way in which they have developed, and in particular the degree and nature of the links into wider economic structures, both of the local area and of the broader international economy, is very different in the two cases.[6]

The international scene

The first point to be made is that the two science parks have utterly different relations with the international structure of new, high-technology development. If what we are witnessing is the emergence of a new technological paradigm, a new era of capitalist accumulation based on changed technologies and different social relations, and an era which is international in its geographical scope, then Aston and Cambridge science parks occupy very different places within it. To caricature, while CSP is a node within an established international network, and directly linked in to that system, Aston is not.

There are a number of indices which, when taken together, demonstrate this difference. First, at the time of our interviews a far higher percentage of firms on Aston Science Park were small independent companies (88.5 per cent) than on Cambridge (54.2 per cent), and in this the two lie on either side of the average found in the UKSPA survey of all science parks (73.8 per cent). In other words, almost half the Cambridge firms are subsidiaries, and linked directly through ownership relations into wider spatial structures.[7] This is true of only just over one tenth of the firms surveyed on Aston. This picture is further elaborated by the data given in figure 6.1, which shows the place of the surveyed science-park establishments within the spatial structures of the companies of which they are part. First, the table confirms the far higher percentage of headquarters on Aston. Second, of those headquarters which were not on the science park, a higher proportion were outside the UK in the Cambridge than in the Aston case. Out of the four subsidiaries on Aston only one was a subsidiary of a foreign multinational; of the ten subsidiaries surveyed on Cambridge, six were part of a group headquartered outside the UK. These differences have continued. In 1990 a maximum of 53 per cent of establishments on Cambridge were independent companies. Of thirty-eight subsidiaries, twenty-two were part of groups headquartered outside the UK.

By 1990, however, the number of independent firms on Aston had fallen, but of the subsidiaries (around 40 per cent) only a quarter were headquartered outside the UK. Another quarter were subsidiaries of local public and economic development agencies, including the West Midlands Enterprise Board (WMEB). A third quarter were owned by or were part of locally based small and medium-sized companies.

These are direct links by ownership. But the picture of more direct international links on Cambridge is reinforced by our data on the location of customers and suppliers. Taking customers first, of the fifteen establishments on each park for which we had detailed data, more firms on Cambridge (nine) than on Aston (seven) exported at least some of their product and, of the

Figure 6.1 The spatial structures of firms with activities on science parks: a comparison of Aston and Cambridge

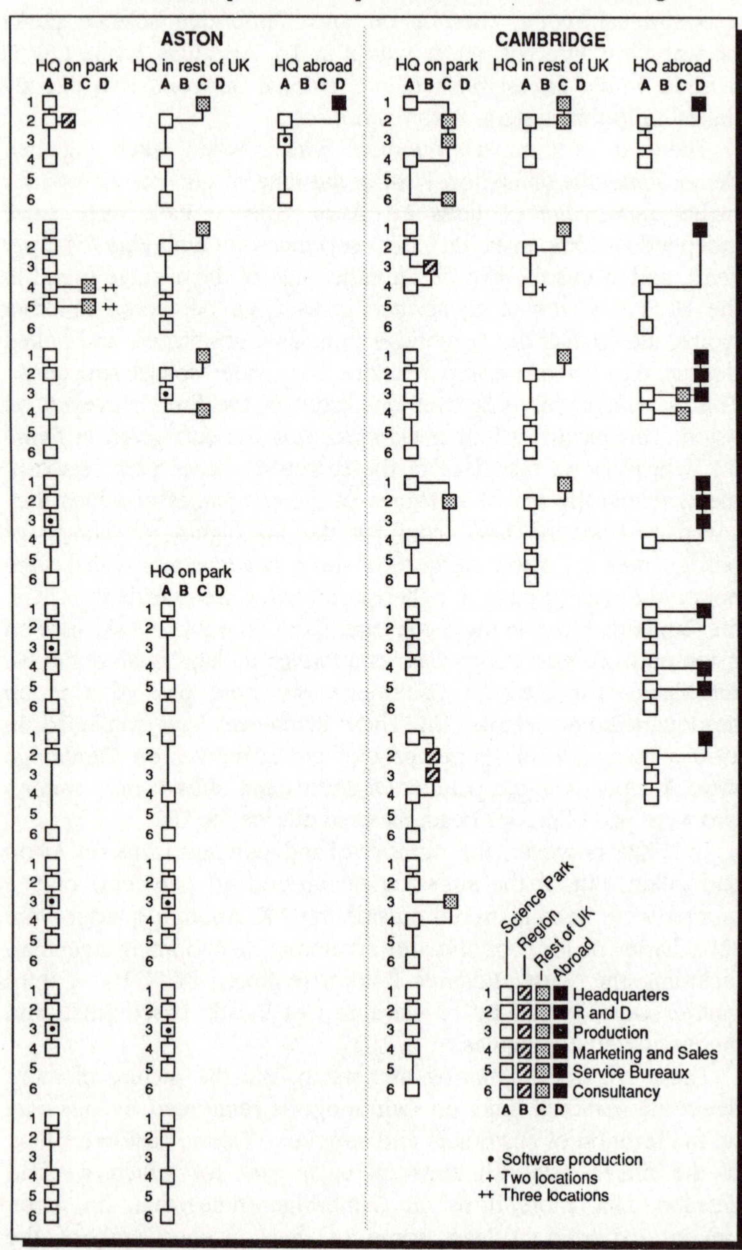

exporters, the percentage of their product which was sold abroad was more than twice as high on Cambridge than on Aston Science Park (an average of 56 per cent as compared with an average of 22 per cent). Only two Aston establishments exported more than 25 per cent of their output, whereas seven Cambridge establishments exported half or more of theirs. The data for suppliers are based on fewer firms, but the overall pattern is the same. Cambridge Science Park establishments, in other words, not only had a higher propensity to export than those on Aston, they also had a higher propensity to import. In this case the averages were 34 per cent (Cambridge) and 24 per cent (Aston). So Cambridge Science Park establishments are more directly locked in, through both ownership and trade, to the international networks of the new technology, the new round of investment.[8]

They are also more technologically sophisticated than those on Aston Science Park. Their activities in information technology are more advanced; as was seen in chapter 2, the range is also wider – there is a notable presence of biotechnology.[9] Moreover, as part of our detailed questionnaire we disaggregated the 'R&D' activity of the surveyed firms into eight sub-activities, ranging from 'doing research' through 'design and development' and 'product development' to 'testing/analysis'. The results confirm that, as a group, Cambridge establishments do more research and Aston establishments more development. Although some Aston establishments are developing relatively sophisticated products, as a group the Cambridge establishments are closer to the 'beginning' of the linear model. These characteristics are further evidenced by interviewees' self-perception of the place of their establishment in the competitive market place. When interviewees were questioned about the criteria on which they felt their establishment competed most successfully, 'reputation and overall image of the firm' was mentioned far more frequently by Cambridge respondents, while those on Aston were overwhelmingly more likely to mention price. Price competitiveness is more likely to be important in selling a more standard product.

It is important, however, to explore what it means to say that Cambridge Science Park is a more internationally connected and more sophisticated place than is the science park at Aston. For one thing, the establishments on Cambridge Science Park are only a small part of the overall high-technology presence in the Cambridge area. The data we have been discussing relate to science-park establishments only. Indeed, other research on the wider 'Cambridge Phenomenon' shows the overall picture in the area to be in some ways very different. In particular, in the area as a whole most high-technology firms (75 per cent) are small, indigenous and independent, while only 12 per cent are subsidiaries of foreign companies (Segal Quince 1985: 27), though they share with the science-park establishments a relatively high (and similarly varied) propensity to export (Segal Quince 1985: 37).

This raises two important issues: the role of the science park in the wider locality, and change over time. There has clearly been considerable change over time, both in the nature of the wider 'phenomenon' and in the fate of individual firms. While small, independent companies clearly remain central to the dynamic of economic growth in the area, there is a growing presence of subsidiaries of national and international companies (Segal 1985). Moreover, a number of originally independent companies have been acquired by 'substantial international (foreign not national) firms' (Segal 1985: 565). Both these shifts relate to the science park. Some of the establishments on the park which are now subsidiaries have indeed been the subject of take-overs, often precisely as a result of their relative technological sophistication. But the higher proportion of subsidiaries on the park than elsewhere relates to another argument – the role of the science park within the Cambridge area. It is a place for established companies, 'offering a high quality property environment in all respects to sizeable, prestige projects of international companies and to substantial local firms with equally demanding property requirements' (Segal Quince 1985: 58). Rents are far higher than for industrial property elsewhere in the area, being rather closer to the levels typical for offices. A number of our interviewees made the point that this is not the kind of place for a risky start-up or a company even remotely on a shoe string.[10] You have to pay for image. It is no accident that the park has a higher presence of subsidiaries of multinationals than the area of which it is part, for it 'has become a visible and prestigious symbol to the outside world that high technology industry is flourishing in Cambridge' (Segal Quince 1985: 58).

Now, if indeed this is the case, one function that the science park is performing is to facilitate a particular form of integration of the Cambridge Phenomenon into the international division of labour. More specifically, it is providing a site for its insertion into the structures of multinational capital, whether by take-over or through the location in the area of its subsidiaries.[11] To be a subsidiary is, however, to have a subordinate status within the international division of labour. This is clearly not a location for subsidiaries in the sense of the assembly-only branch-plant economies of some northern parts of the UK. 'Branch plant' and 'subsidiary' can mean many things. What is at issue is their place within the overall geography of the relations of production (Massey 1984). None the less, non-HQ status means that ultimate control and power of decision-making over the accumulation process as a whole lie elsewhere and this in turn may mean that other functions get down-graded relatively. When we interviewed them, two of the big foreign-owned multinational subsidiaries in Cambridge were using their science-park facilities for marketing and sales (see figure 6.1); their research and development was in their home countries – the Netherlands and the USA respectively. The Cambridge Science Park and the wider phenomenon may be more linked into the currently leading-edge international networks than is Aston, but they are not at its forefront.

There are fascinating contrasts here. The Cambridge area has moved from being a sleepy backwater (in investment terms) which participated hardly at all in previous major eras of manufacturing accumulation to being one of the key links (perhaps *the* key link) from the UK economy into the current era. Yet it is a second-rung link. In this major round of investment UK-based capital no longer leads the world. This is a trajectory utterly contrary to the one which Birmingham has traversed. That city and region were for decades in previous rounds of accumulation one of the leading pivotal focuses of the international economy, the home base of multinationalised manufacturing capital and in a position commanding much world trade. The Cambridge area today cannot emulate that position. But as we saw earlier it was precisely some of the inheritance of that previously commanding position that had left the Birmingham area today with characteristics deemed on the whole to be inimical to the current social form of high-technology development.

In the comparison pursued here, it is Aston Science Park which today has the new start-ups, the independent firms and the 'entrepreneurship'. There has indeed been a deliberate effort precisely to attract on to the park a 'flagship' company: 'one or two well known "flag-ship" companies would be particularly welcome', says the

1985 EDC Science Park Working Party report 'whether their presence is directly associated with research and development or not. They might usefully be located towards the periphery of the Park without significantly affecting land availability for "purer" Science Park users' (para. 4). And it is suggested that financial incentives, perhaps in the form of rent relief, could be used to attract on to the park 'more established firms (perhaps moving from elsewhere) who don't require venture capital support' (para. 6ii). And indeed, nicely resonant of the previous era in its use of this land, Delta has located a facility on the science park.

What these contrasts reflect, of course, is not only the parks' different places in the international division of labour but also their contrasting relations to the local economy, and to local economic change.

Links with the local economy

If Cambridge Science Park is more directly linked into the international economy, it is the establishments on Aston which have the clearer links with other firms in their local area. A glance at the literature on industrial organisation and the geography of multiplier effects and inter-firm linkages might imply that this is almost definitionally the case. The overwhelming general conclusion in that literature is that small, independent firms are more likely to have local linkages (though, as is less frequently recognised, these linkages are likely to be correspondingly small) than are the subsidiaries of major multi-regional or multinational companies. Aston, as has already been shown, has a far higher proportion of small, independent firms on its science park than does Cambridge. It might seem to follow therefore that this is 'the explanation' for the higher levels of linkage from that science park to its local economy. In fact, however, the argument is more complicated, and more interesting, than that.

First, the evidence from Segal Quince (1985) on the Cambridge Phenomenon as a whole neatly upsets the current consensus. Cambridge firms are not only overwhelmingly small and independent but also 'as a whole depend little on local input and output links: 74% and 72% respectively reported no or minor links of each kind' (p. 37), though the research was unable to assemble systematic quantitative information on sales and purchases for the sample as a whole. There may in the past have been a correlation between size and the geography of linkages, but such taxonomic groupings provide neither the basis of a real explanation (Sayer 1984) nor an adequate account of the character of capital (Massey 1984). As Segal Quince say, reflecting on their results, 'This [low level of local linkage] is not surprising; one would expect most high technology firms to have to operate in much wider market-places' (p. 37). This point is fully borne out by our own results, for while Aston has a somewhat higher level of linkage to the local economy than does Cambridge, in both cases the levels of local multipliers are in fact very low.

As with the study of the broader Cambridge Phenomenon, our evidence on science-park firms concerning the intra-UK distribution of customers and suppliers is limited and uncertain (it is amazing how many dynamic entrepreneurs have very little idea of what is going on). However, the precise company-level data we did gather indicated similar proportions of customers (around one-third) within fifty miles of each park, but a slightly higher interaction between Aston establishments and their immediately local area (up to ten miles), than in the case of Cambridge, especially in the case of suppliers. The Aston data can be amplified from the 1990 local authority survey of forty-six units. This underscores the low level of multiplier linkages, even on Aston, into the local economy. Only 10 per cent of key markets were found to be in Birmingham, with a further 15.2 per cent in the West Midlands; and of suppliers 7 per cent were in the West Midlands beyond Birmingham, and 23 per cent in the city itself. The latter, however, consisted mainly of low-tech consumables (for instance, office stationery); specialist equipment was generally purchased outside the local economy (Birmingham City Council 1990). These figures make clear the importance of distinguishing between types of links. Indeed, the fact that most of the sales of Aston establishments are outside the region can be seen as a definite success in terms of the aim of establishing new basic (i.e. regional exporting) sectors.

This picture is amplified by our detailed data on links between science-park establishments and large companies generally. The only category of link on which Cambridge establishments registered a significantly higher response than Aston was, as we have already noted, that they were much more likely to be owned by a larger group. But Aston companies had three types of links with larger firms which were much more strongly developed than in the Cambridge case. These were (i) that they made use of the expertise of a key person, (ii) that a large firm was the source of a product or of market strategy, and (iii) that product development took place in a large company. (Neither 'management buy-out' nor 'large firm as key customer' showed up as being important on either science park.) The nature of the Aston links hints at the possibility of personal and quite intimate connections. Moreover, it is notable that a high proportion of the links are with major firms which are either based in the area or have a substantial presence there. This was not true in the case of the, anyway fewer, links from Cambridge Science Park firms to large companies. We followed these Aston links up, and discovered a fairly dense set of connections between a number of Aston Science Park establishments and some of the major regional companies. Some of the old dominant West Midlands firms – Lucas, Delta and Dunlop in particular – showed up in this structure of interconnections.

The links between Aston Science Park establishments and other local companies, then, are quite particular and are different from those pertaining in Cambridge. Part of what this reflects is the slightly different set of functions performed by the establishments on the two parks. More of the Aston establishments are 'service' firms, consultancies and suchlike, performing functions which are interstitial within the dominant structures of the regional economy. Aston establishments derive a higher proportion of their income from training, and a slightly lower proportion from the sale of physical products; figure 6.1 shows the higher level of activity of Aston establishments in service-bureau functions and in consultancy. A second, though probably minor, element is a fairly conscious attempt to extend to the present round of investment the long tradition in the Birmingham economic structure of relations between large and small firms. The 1985/6 Economic Strategy for Birmingham, for instance, mentions a few times the importance of building on this tradition of linkage (Shaylor 1985: paras 3.8, 3.12–13); and in interviews with management of Aston Science Park the tradition of subcontracting within the Birmingham economy was stressed as a strength. Third, there is a wider point that a number of Aston establishments are linked to local economic strategies. Thus there is the Advanced Manufacturing Centre, a joint venture between Birmingham City Council's Economic Development Unit and Deltacam Systems, the Environmental Protection Unit laboratories, part of BCC's Environmental

Services Department, Information Technology and Expert Systems Unit (run by Birmingham Education Authority's Continuing Education Division), and West Midlands Technology Transfer Centre, a joint initiative of West Midlands Enterprise Board and Aston University to promote technology transfer into West Midlands industry. None of this should be over-emphasised. The degree of linkage by Aston establishments into the local economy is not large; it is simply greater than at Cambridge. And there were comments, too, in interviews which pointed in the opposite direction. One firm which is precisely in the business of applying new technology to manufacturing industry, and which is in a key sector (metals) in the local economy, none the less had only 15 per cent of its customers within the West Midlands (in a radius of fifty miles). And another stoutly maintained that it was impossible to find suppliers of metal components locally (in Birmingham!) and that it was necessary, for both price and quality, to seek them outside the region.

Labour markets

But it is in relation to labour and labour markets that the real paradoxes and contradictions of science-park strategies arise, and where the issues link back clearly to those of previous chapters. It was seen in the second section of this chapter that in neither Birmingham nor Cambridge was job creation in itself the main immediate dominant objective. However, in both areas the city council had recognised serious imbalances in the supply of and demand for different types of skills, which their own economic strategies as a whole were supposedly geared in part to ameliorate. In both cases the concern was over the lack of jobs, and the low wages, for certain workers, particularly those not classified as skilled. There is no evidence in either case that the science-park strategy in itself has done anything, or indeed could do anything, to improve this situation.

In 1986 the review of Birmingham's economic strategy (*Birmingham: the Business City*) continued to point to a growing polarisation in the labour force, the emergence of a 'two-tier' labour market. It noted a steady growth in low-paid employment, a significant widening of income differentials, an expansion of employment (such as part-time jobs) without the range of legal entitlements or security, and 'a slowdown, or standstill, in progress towards equal opportunity for ethnic minorities, women and people with disabilities' (section 3). It points out that such changes, which are happening at a national level, are likely to have a particularly adverse impact on an inner metropolitan authority like Birmingham. 'For many Birmingham people, the prospects for their working life are alternating bouts of unemployment and very poor quality work' (para. 3.5). This picture is reinforced by information on vacancies and unemployment. There has indeed in the later 1980s, in line with national trends, been an increase in notified vacancies in the city. 'However, this growth was concentrated in "professional" and "service selling" occupations. Notified vacancies in the "construction" and "miscellaneous" occupations – those most suited to the skills of those workers who form the bulk of the registered unemployed – actually declined in numbers. This mismatch between the skills and jobs available in the City is highlighted by the fact that, despite the high levels of unemployment locally, skills shortages are reported in engineering trades and new technology trades' (paras 2.7, 2.8).

The profile of employment on Aston Science Park shows that over 60 per cent of those employed are qualified scientists and engineers. This is a higher percentage, indeed, than in Cambridge, where the presence of some larger firms shifts the balance between occupational categories. But Aston is also way over the national average for science parks in the proportion of its establishments which have over 80 per cent, over 60 per cent and over 40 per cent of their employees in the QSE category (UKSPA-OU-CURDS). Of these, relatively few are women or members of ethnic minorities.[12] Of course, this kind of occupational structure reflects the very definition of a science park as a place where research and development is separated off from production. It also reflects the national characteristics of these high-status employees. Although, as we saw in chapters 2 and 3, science parks as a whole do not quite live up to their ivory-tower-plus-commercialism images in terms of occupational structure, they are none the less obviously not sources of jobs for manual workers, skilled or unskilled, or for women or for ethnic minorities ... This is recognised in city council documents. The need to spread access to technology is mentioned in reviews of the economic strategy and in a review of the science park itself, in which it *is* claimed that job creation is a major aim (it comes second, after 'widening of the City's economic and industrial base'), it 'is accepted that new technology industries may not create as many jobs as more traditional industry in the past and that job creation has to be viewed over a fairly long timescale, with significant numbers coming from the second-stage expansion of newer companies – in a 5 to 10 year timescale' (EDC, 26 June 1985). Even if this view of the medium term is accepted as credible, the issue is not just numbers of jobs but the types of jobs and who does them.

The polarised and élitist occupational and social structure of new high-technology sectors is a general phenomenon, although arguably more marked, as we have seen, in the UK than in some other countries. This poses serious dilemmas for any local council in relating new technology to any meaningful local economic strategy. A 'classic' science-park strategy in itself, without very major countervailing social policies and/or a strong belief in the generally discredited theories of 'trickle down', seems likely to reinforce tendencies towards polarisation within labour markets. It does this in three ways. First, and as was spelled out in some detail in chapter 3, it is predicated upon a model of innovation which is inherently élitist, which stresses the social separation between those who think and those who do, between conception and execution. Second, as spelled out in the last chapter, the whole point of a science park is physically to separate those who think from those who do – indeed, to group the commercial conceptualisers with those of academe. Spatial separation thus reinforces social separation. Third, and again as discussed in chapter 5, science parks are designed as élite places. And it is not possible to have élite status without those who fall outside it. It is a necessarily positional concept, presupposing, and through its rhetoric attempting to reinforce, inequality.

There have been many instances in recent years, perhaps particularly in inner cities, of new developments being resisted by local people. Either the employment to be generated is not for them (it is too high-status, skilled, etc.) or the jobs on offer are only boring, low-status and low-skilled. (Thus the 1990 survey on Aston – where there has been a consciousness of the issues involved – found that of a total of 704 employees in the establishments surveyed 51 per cent were resident outside Birmingham and an additional 26 per cent were resident in the outer city. Twenty-three per cent lived in the inner city, where the science park is located.) The problems in such cases are not fundamentally those of geographical maldistribution. At heart they relate back to the very division of labour in society. And in high-technology industries in particular, as we have shown, this in turn relates back to models of science and innovation. Yet, as we have also seen, in chapter 5, geographical form has its own implications for the social content and meaning of the division of labour. Through their insistence on the spatial separation of stages in the division of labour and the importance of their spatial imagery, science parks are once again a particularly acute exemplification of a much more general social phenomenon.

The Cambridge case provides a different kind of evidence for this. As we have seen, the history of the Cambridge local area, far more than that of Birmingham, had been one of polarisation: the juxtaposition of an élite with a relatively unorganised, and very low-paid, working class. This, of course, was not a concern of Trinity College, nor of the university, in the establishment of the science park. None the less, as we saw, the initiative was backed by the city council. A comparison of the high-tech sector with conventional manufacturing in the Cambridge area as a whole is indicative of the results. First, 'the survey indicates that there were more job opportunities for white collar staff in the high tech sector while in the conventional sector there were greater opportunities for blue collar employees. ... blue collar staff accounted for just 22% of jobs in the high tech sector as against 60% in the conventional sector ... there were relatively fewer job-opportunities for the semi and unskilled in high tech companies compared to conventional ones. For this group, which is of key interest to the Council, the proportion of jobs in high tech fell to 11% as against 26% in the conventional sector' (Cambridge City Council 1986: para. 3.09). The other outstanding contrasts were for skilled manual workers (34 per cent in the conventional sector compared with 11 per cent in high tech) and scientific/technical occupations (12 per cent as against 47 per cent). The report makes a number of points about these results: first, that if a distinction is made within high tech between research and development and manufacturing companies, the latter emerge as less different from the conventional sector, and second that the bias of high-tech companies lessens with age. The second point is open to question, since it is based on static data (a comparison of firms of different ages) while the conclusion drawn refers to a process (the ageing of particular firms over time). However, the first point stands. What it does, however, is confirm all the survey results showing an even more élitist occupational structure on science parks, where the bias is more likely to be towards research and development.

Of course, that would not be important, in these terms, if the manufacturing associated with the research and development were elsewhere in the area, even if off the park. In fact, however, 'such large scale production as exists is typically subcontracted elsewhere' – outside Cambridge (Segal 1985: 564) the companies 'are unlikely to be allowed to grow very big within the immediate locality (the Cambridge area), because both city and county are looking for them to be seed-bed, research-based concerns which will find another site once manufacturing becomes important' (Moreton 1986). Land and property prices anyway pose problems for larger-scale production, but there also continue to be planning restrictions on industry in the area. All this fits well, of course, with the university's aims in allowing a (highly specific) form of development in Cambridge. 'Cambridge Science Park has come to constitute a no less significant symbol to the "inside" world (the University) of the success of Cambridge as a location for high technology industry and of' – and this is the real point – 'the easy compatibility ... of this industry with Cambridge as a university town' (Segal Quince 1985: 58). Compatible with the university's views it may be; but it does not do much for the low-paid or less-skilled.

None the less, the overall growth of the city's economy led to unemployment rates overall of below 3.0 per cent at the turn of the decade. The chief problem from the city council's point of view became that of upgrading employment rather than increasing it. However, the wage distribution within high tech reinforces the problem for the semi-skilled and unskilled manual workers. Table 6.1 shows the average wages and salaries in the high-tech and conventional sectors. The percentages show the greater polarisation within the former sector. Higher salaries are in general higher and lower salaries lower in the high-tech sector.

Other aspects of this increasingly polarised labour market were also revealed in our detailed survey interviews. Overall the high-tech sector in Cambridge has a higher proportion of women (28 per cent!) than does the conventional manufacturing sector (22 per cent). Most of them, however, are confined to clerical work. It will be noted, moreover, from table 6.1 that clerical workers fare rather better in high-tech industries, on average, than they do in the conventional sector. In the conventional sector they are the lowest paid. (The ordering of the occupational categories down the page in table 6.1 is the one used in the report cited, and indeed the normal one, and reflects the non-manual/manual division rather than the pay hierarchy.) In the high-tech sector their pay is higher both absolutely and relatively, and may reflect the growth of a designer secretary/receptionist labour market in keeping with the general ambience.

Table 6.1 Average wage and salary levels in Cambridge (£000s per annum)

Occupation	*(a) High-tech average*	*(b) Conventional average*	*(a) as % of (b)*
Managerial	18.0	12.8	140.6
Scientific/technical	13.4	10.3	130.1
Clerical	7.3	6.1	119.7
Skilled manual	8.6	8.4	102.4
Semi-skilled manual	5.5	6.5	84.6

Source: Cambridge City Council (1986: para. 3.12)

But, as we saw in chapter 3, the expected dominance of (white) men in high-status jobs was confirmed. In one of our interviews, discussing research into the training needs of new-technology firms in the Cambridge area, it was commented that there *is* a role for women as managers 'especially part-time in small firms' but firms are loath to advertise for part-time managers 'because they don't want second-rate'. The question therefore arises as to the nature of the labour market for graduate women in Cambridge. A hint was given in interviews with firms. When one of them advertised for a receptionist 'several girls [*sic*] with degrees applied'. The possibility that the UK's new high-tech regions are generating particularly sexist labour markets has been commented on in other research (Boddy *et al.* 1986; Massey 1988).

Of course, as we have said, science parks in themselves cannot be expected to solve all the problems of local labour markets. They are seedbeds for future growth, etc. Our point, however, is that through their élitism and through their very definition in terms of social-spatial separation they exacerbate the anyway growing polarisation. Segal Quince (1985), with an insouciance which (surely?) belies what they really know, point to 'the low proportion of "ordinary" jobs among the newer firms, as would be expected. But because of high profits and high value added per employee, the firms generate above-average employment multiplier effects especially given the historically low wage levels in the local economy. The readily evident expansion and increasing sophistication of the retail, recreation and business and personal services sectors currently taking place must in good measure be due to the growth of local high technology industry; and these sectors mostly offer "ordinary" employment' (p. 35). The question, of course, is what 'ordinary' employment is like and how much it gets paid. What may well develop, as Boddy *et al.* (1986) show in their study of Bristol, is a division between a core of highly paid white males in professional and technical jobs and a servicing sector of low-paid, often casualised work. The multiplier effects referred to by Segal Quince (1985), it will be noticed, are almost all income multipliers. (We have seen earlier in this chapter that the technological multipliers are relatively low.) These jobs, in other words, depend on the high incomes of others. Such multiplier effects are truly the crumbs from High Table.

The problems, of course, do not end in the labour market. House prices have risen fast and 'the lowly-paid, of whom there are many in Cambridge working in college domestic jobs or part-time tourist work, are effectively squeezed out of the housing market' (David 1986). Indeed, the pressures of development have given rise to a number of conflicts in the Cambridge area. High-tech strategies as represented in science parks can serve to exacerbate what are anyway increasing occupational inequalities.

There is, moreover, a further issue. It was seen in chapter 5 that there are severe national shortages of scientific and technical labour in the UK. (That, indeed, is part of what gives these strata their monopoly power and privileged position.) But there is a definite geography to these shortages, which has been documented at regional level (see chapter 5) and which shows up clearly in the contrast between our two detailed studies. In the UKSPA-OU-CURDS survey, 24.5 per cent of establishments interviewed reported shortages of skilled labour in their first year, and 75.5 per cent no such shortages. On both Aston and Cambridge a higher level of shortage was reported than nationally. Given the analysis in this chapter so far, however, it may be hypothesised that the balance of reasons for these shortages is different between the two areas. In Cambridge, as was seen in chapter 5, the shortages arise in a 'virtuous circle' of influx of both jobs and people, with companies continuing to go to the area because the very size of the high-tech labour market holds out the promise that any individual enterprise will be able to acquire the labour it needs even in a wider context of overall shortage in the area. In Aston the shortages are more likely also to result from a simple lack (in the sense of low proportion) of such skills in the local population of the inner area of a major conurbation. Moreover, while on Aston the figure was 30.8 per cent that for Cambridge was 40.0 per cent. In other words the problem was reported to be more severe in Cambridge than in Aston. There was also difference in detail. On shortages both of 'skill/experience/knowledge' and of 'specialists' more shortage was reported in Cambridge than in Aston. For the general 'shortage of suitable people' there was more of a problem at Aston, but while 25 per cent of Cambridge respondents reporting a shortage said it was 'always a problem' not a single Aston respondent said this. It also seemed to be clear that the geography of the labour market for technical and scientific workers was wider (including more international) at Cambridge than at Aston (and wider than for blue-collar workers – see table 6.2). Indeed, and significantly, in Cambridge location on the science park was mentioned occasionally as a way of attracting top-class labour. This is a significant point. Given what we have already learned about rent levels, it means that the shortage and the bargaining power of these workers puts up costs to the companies through location costs as well as more directly through salaries.

Table 6.2 Locally recruited personnel, Cambridge

Occupation	*%*	*% recruited locally*
Managerial	23	24
Scientific/technical	37	28
Administrative	19	75
Skilled manual	6	59
Semi-skilled manual	9	64
Unskilled manual	6	85

Source: Moore and Spires (1983: table 4)

Research links

The contrast between Aston and Cambridge in the nature and level of research links reinforces the arguments already put forward. Establishments were asked to specify the nature of up to three of their most important types of links with the local institute of higher education. There were clear differences between the two parks. Overall, and perhaps surprisingly at first sight, there was a higher response rate from Aston establishments. However, the types of links specified were very different between the parks. A far higher proportion of the Aston responses concerned the use of general facilities (library, recreation, common room, etc.) than the area of technology transfer or links through individuals. The use of university facilities was both absolutely much higher, and relatively more important, at Aston than at Cambridge. One thing which this may illustrate is the relative shortage of facilities in its inner-city location. But Aston Science Park personnel have also made efforts to construct links, and this may help account for the overall higher level of formal contact than in Cambridge. Thus local academics take part in the appraisal system for new tenants, there is a specific appointment to facilitate use of the university library, and so forth. However, evidence from a very recent survey indicates that the relationship between the university and the park may have changed with time; the initial impetus for the high level of contact does not seem to have been fully maintained (Birmingham City Council 1990)

For establishments on Cambridge Science Park it is research and development and personnel-related links which are more important, though even here the absolute level of contact is not high and is lower than at Aston. Moreover, further disaggregation again shows the balance of components to be very different between the two locations. In Aston there was a considerable spread of types of responses, including in particular informal contact with academics, access to specialist equipment, student project work and employment of recent graduates. In Cambridge the overwhelmingly dominant type of link referred to was informal contact with academics. This informality of the links at Cambridge is important. In some of our interviews it was indeed stressed that other research about Cambridge Science Park (such as that by Moore and Spires 1983) had underestimated the level of links between the park and the university precisely because they had looked for formal links, and asked the 'wrong' (i.e. too narrow) questions. This point is certainly reinforced by the results of our survey. And the distinct contrast with the types of links at Aston underlines its importance. What is at issue is a cultural and political difference. At Aston there is a real attempt to set up working links and to use the spatial juxtaposition of the park and the university to counter some of the perceived disadvantages of the locality and of the institute. In Cambridge it is much more a question of an informal set of networks, an internally referential clique. While this may in some sense 'work' it is based on élitism, is probably difficult for outsiders to penetrate, and is not simply reproducible elsewhere. It is, moreover, a research community within the Cambridge area which is wider than the university and the science park. We have already seen that Cambridge is the location for a number of important national research laboratories and, as we shall see in the next chapter, there is a significant number of important contacts between them and establishments on the science park.

OTHER CONTRASTS

The cities of Birmingham and Cambridge stand in very different relations to the history of geographical uneven development within the United Kingdom. The character of their science parks reflects this and the parks play highly contrasting roles within their local areas. Cambridge Science Park has caught the moment; it is part of the emergence of a new era of accumulation, though it plays a specific role within that process. Aston Science Park is one element in a wider attempt to escape from the inheritance of previous eras.

These are just two examples; there are others. Indeed, Birmingham City Council itself in 1986 established a second science park which was designed to be different from, but complementary to, Aston. This is Birmingham Research Park and it is related to Birmingham University. Here the emphasis is very much on companies emerging from research at the university. Birmingham R&D Ltd has been set up to handle the transfer from university to park, and it retains intellectual property rights. Moreover, within the West Midlands conurbation there is a further contrasting view of how to proceed, given the region's industrial history. The strategy of the West Midlands Enterprise Board contrasts with that of Birmingham City Council as embodied in Aston Science Park. The problem is what should be the relation to 'new technology' in a region which lies outside the main centres of its growth, and indeed whose industrial history in every way, physically and socially, seems unlikely to attract such growth, given its current social form. The West Midlands Enterprise Board strategists would see Aston Science Park as geared to new start-ups in non-traditional (for the West Midlands) sectors, and to promoting innovation (interviews). In other words, it is an attempt to generate new export bases with no necessary relation to the industrial history of the region. As we have seen, this is in fact only one role of Aston Science Park, according to the various documents of Birmingham City Council; the park also aims to have a role in modernising existing basic industries. Moreover, as we have also seen from the brief analysis here, it is this second role which, so far, would seem to have the greater potential.

It is this latter approach to local economic regeneration which is stressed by the West Midlands Enterprise Board. It argues that innovation is not centrally important, indeed that it is an economically insignificant red herring, and that the regeneration of the West Midlands economy will come not from new firms but from the revitalisation of existing ones. Its strategy, therefore, is principally geared to the diffusion/application of new technology into existing industries. Within this there is a particular focus on components sectors, prioritising 'investments which can make the most strategically significant contribution to rebuilding linkages between West Midlands companies and sectors' (Marshall 1987). It is into this strategy that Warwick Science Park (in Coventry) has been incorporated. Both the value and the possibility of new-product innovation are questioned, the latter particularly in the context of the difficulties faced by new start-up firms innovating and trying to find new markets. Much more important as a problem, and more likely to be effective if tackled, it is argued, is the lack of investment in new process technologies. The focus, in other words, should be on diffusion and process technology, rather than on innovation and new products.

At its clearest, the debate is between trying to get in on new technology as a sector in itself and trying to use new technology as a means of revitalising the existing economic base. Both approaches clearly face huge obstacles; and in both cases this is tied up with the nature of capitalist accumulation and competition on the one hand and the fact of geographical uneven development on the other. The problem faced by the first strategy (trying to become a regional base for new technology sectors) is clear and has already been discussed: Birmingham is not Cambridge and it is precisely Birmingham's history and indeed its very need for jobs (the fact that it is an area with serious economic problems) which make it unattractive. The area has already been 'used' by manufacturing capital, and the social and physical characteristics which are partly a product of that period of exploitation now make it unsuitable for further investment. The challenge here is to counter the force of capitalist spatial uneven development.

The challenge to the strategy of revitalising the area's existing economic base is different. Here the argument is that not all the growth must be with the new: 'The choice so often presented by Government ministers between supporting older, declining "sunset" industries or fostering new, expanding "sunrise" sectors is, in many cases, a false one' (Marshall 1985: 576). It also necessitates arguing that the decline currently experienced in the local economic base is not, at least not fully, 'necessary'. That is to say, there are reasons, and reasons which are remediable, which mean that the decline thus far is worse than it might have been.[13] Thus the West Midlands Enterprise Board strategy of which Warwick Science Park is part starts from such an analysis:

> The overdependence of the West Midlands economy on nationally declining manufacturing industries, however, does not provide a complete explanation for the region's economic plight. In many instances, West Midlands manufacturing industries have not merely contracted in line with national trends but have declined at a faster rate than the same industries at a national scale. The West Midlands has suffered particular problems over and above the wider difficulties experienced by British manufacturing industry.
>
> (Marshall 1985: 571)

The key here is the West Midlands' long history of chronic lack of investment, and especially investment in new process technologies, a phenomenon which has been widely documented (Massey and Meegan 1979). The resulting technological backwardness of much West Midlands manufacturing 'reflects the activities of giant multinational companies which command a substantial proportion of the local economy' (p. 571) and 'in part at least, the failure of past governments to develop a comprehensive strategy for industrial investment' (p. 570). Thus, it is argued, it is possible to challenge the existing dynamics of capitalist competition – there is some room for manoeuvre. Taking advantage of that room for manoeuvre, however, it is further argued, cannot be left to the dynamics of capitalist competition:

> A far-reaching renewal of the West Midlands' industrial base, however, will not occur spontaneously. Past experience suggests that, left to its own devices, the private sector will be unable to achieve the levels and types of investment in fixed capital, skill training and new products or production technologies to revitalise the regional economy. There needs to be an expanding role for public sector intervention to support economic regeneration in the region.
>
> (Marshall 1985: 576)

Warwick Science Park (formed and funded by Warwick University, the West Midlands and Warwickshire County Councils, Coventry City Council and Barclays Bank) is somewhat specific within the above strategy in that it is indeed focused on innovation. In its sectoral remit, however, the aim is to link into the region's existing industrial base. It 'is intended to assist the generation of *new* innovations by forging stronger links between manufacturers' own research and development resources and the region's established scientific community' (Marshall 1985: 574). It is focused on CADCAM, FMS and robotics; on sectors, in other words (electronic or computer-based manufacturing systems, mechanical and electrical engineering), with immediate relevance to the technological needs of the regional economy. The 'needs of the local economy', of course, tends to mean the needs of companies. It is not for nothing that the local HEI has been called 'Warwick University Limited' (Thompson 1970).

The implicit debate about strategy for Birmingham and the West Midlands, then, touches on a wider debate about how regions not currently at the centre of accumulation in the new era can best link into 'new technology'. And it is a debate reflected in the contrasting strategies of the respective science parks.

For a final example, take the case of east central Scotland. The wider region in which Heriot-Watt Research Park is located (that of central Scotland) has had previous rounds of investment which are in longer-term decline than those of Birmingham. However, the position of Edinburgh itself, with its university, its function as administrative capital, its low historical level of manufacturing and its position as an important, and growing, financial centre should provide tangible and intangible advantages for the location of 'high tech'. And indeed it has. In the mid-1980s, 40,000 people were employed in 'electronics' in Scotland, the term 'Silicon Glen' has entered common parlance, and the Edinburgh area in particular has strengths in precision engineering, electronics, medical instrumentation and biotechnology. Ironic evidence for this has already been mentioned in chapter 5, where it was pointed out that Edinburgh has an intra-regionally high level of shortages of scientists and technologists. This is an area, then, with a base in high-tech industry far more developed than in the area around Aston, if also far less than in the area around Cambridge.

But it is also a development of a different *type* from that around Cambridge. Among the many stimuli to its growth (particular departments at Edinburgh University, the work of Ferranti) has most recently been the activity of the Scottish Development Agency. After early debates, the dominant strategy adopted by the SDA for the regeneration of the Scottish economy was, first, to prioritise high tech as a new sector in preference to any attempt to modernise existing sectors and, second, to produce a high-tech sector through attracting inward investment.[14]

The first of these choices reflects the debate in Birmingham/West Midlands, with the SDA coming down on the same side as Aston – the need to develop 'high tech' locally as a basic sector in itself. Aston's strategy is to 'grow' that sector locally, and the difficulties of doing so have already been hinted at. The SDA strategy, in contrast, was to attract in an 'already developed' sector. That strategy, too, however, has its difficulties. Certainly there has been job growth (though tiny in comparison with the numbers once employed in the so-called 'traditional' sectors). But it is, overwhelmingly, a branch-plant economy. There is little local ownership. This, in turn, is reflected in the functions performed – there is some R&D (Morgan and Sayer 1988), but it is very restricted and far more of the D-type than serious R (interviews). The level of local purchasing (within Scotland) is low (17 per cent of main inputs by value, 12 per cent of the total; Buxton 1987), and here this is a reflection of the level of foreign ownership and the position of these branch plants within international hierarchies of production, as well as of the generally more internationalised nature of high tech. The 1979 Booz Allen & Hamilton study which formed the basis of the SDA's current electronics strategy specified a (low) target of 1,300 jobs to be created in Scottish firms supplying multinational electronics firms. By 1986 barely a hundred new jobs had resulted (Danson *et al.* 1987). More specifically, and reflected in and reflecting all these characteristics, the now quite considerable period of growth of electronics in Scotland has not seen the development of a local components industry. Moreover, the challenge of promoting a new components sector is now increasing as the major firms, in order to improve quality control, move towards closer and longer-term relationships with smaller numbers of suppliers (*Financial Times*, 31 March 1987; and interview). Thus the lack of local purchases is perpetuated and the 'sector' remains a collection of plants rather than developing into an integrated new industrial base.

It was in this context that Heriot-Watt Research Park (HWRP) was established in 1972. It was a context which offered a number of potential roles for such a development. It could have adopted a strategy, as at Warwick, concentrating on the modernisation of existing sectors. This was indeed an important element in the original thinking of the Scottish Development Agency, which was established in 1975. It was in and after 1979, with the change of government from Labour to Conservative, that the strategy of the SDA switched so dramatically to its current direction. Had HWRP opted for a strong policy towards existing industries it would now be operating a strategy to counterbalance that of the SDA. Indeed, it has been argued that such a strategy is even more necessary as a result of the current activities of the SDA, whose policy concentration on 'sunrise sectors' can be argued to be actually further disarticulating the existing Scottish economy (Danson *et al.* 1987). Alternatively, there is a clear and urgent need for an integrating role within the high-technology sector itself, to try to plug some of the gaps in the existing collection of companies. One obvious integrating function might have been an all-out attempt to generate a home-based components sector.

It is difficult to assess what indeed is the role of Heriot-Watt Research Park in the Scottish economy. We had wanted to make it our third major intensive study. However, along with those of other researchers, our enquiries were at the time fended off.

Interviews with companies were denied by park management. What follows is necessarily, therefore, based on highly incomplete information. What it reveals, however, is that HWRP is playing neither of the above potential roles.

In the interviews which we were granted it was stated that this research park does not need to be a catalyst of change in the local economy, since that change is taking place anyway. 'Edinburgh *is* a giant science park; all resources, supplies, infrastructure [are] here. We have precision manufacturers, component manufacturers, etc, etc. Heriot-Watt is successful *in the region* in an integrated way, not isolated from the region. It can only do as well as the surrounding community.' What is more the park does not have the problem of having to create jobs in the local economy, because 'Edinburgh doesn't have an unemployment problem'. (In fact Edinburgh's rate of registered unemployment has moved more or less with the national average – at the time of this statement it was 13 per cent.)

The focus of HWRP strategy is on innovation, commercialisation and technology transfer. Perhaps the most significant element on the park is the set of Technology Transfer Institutes (TTI), which are university-owned but commercially based. Their aim is to 'bridge the gap between possibilities opened up by new technology (wherever it is developed) and existing industry. Existing industry ranges from sophisticated firms able to assimilate leading-edge technology, to unsophisticated firms needing to be put on the learning curve'. Thus far the aims share much with those of WMEB. The difference comes in the selection of sectors. From the information available it does not seem that it is existing sectors of the Scottish economy which are primarily of interest; areas of intervention are not selected in relation to an analysis of economic restructuring. Thus, for instance, a strategy to increase the local components supply to electronics multinationals would involve a range of sectors both 'sunrise' and so-called traditional. It would include, for instance, products such as plastics moulding, metals and electro-mechanical parts (Danson *et al.* 1987). A report prepared for the Scottish Council (Development and Industry) in 1983, which noted the lack of local sourcing by US and English firms in Scotland, identified 'opportunities to improve domestic supply particularly in instrument engineering, shipbuilding and vehicles, metal goods, and paper, printing and publishing' (*Scottish Economic Bulletin,* No. 27, cited in Danson *et al.* 1987).

But Heriot-Watt Research Park has set its sights on a different goal – that of new industries. Given the analysis of 'no local economic problems' 'there are no short-term objectives. For instance, in the future Heriot-Watt University and research park may be able to say we helped get slices of future new industries, such as medical lasers, of which we had the first in Europe' (interview). Similarly, Bioscot Ltd – a collaborative venture in biotechnology – is related to Heriot-Watt Research Park.

A further aspect of this strategy is that there is a stricter than usual regulation against manufacturing production being carried out on the park. Indeed, this is called a research (not science) park, and the appellation 'Science Park' was quite forcefully rejected in interview, in part for the very practical reasons that there was no money to put in the necessary infrastructure, there was no relevant expertise and, on the other hand, there were anyway plenty of industrial sites available near by.

Reflections

Science parks epitomise the kinds of issues facing areas caught up in the shifting structures of uneven development as accumulation moves from one round of investment to another. While Cambridge Science Park is carried along on the area's general growth, can Aston or Heriot-Watt generate a similar round of new investment in their areas? Or is the strength of the locational pulls of 'new technology', as a sector in itself, too great to resist? The evidence here, preliminary as yet but to be expanded upon in the next chapter, is that the struggle will be an uphill one.

There are, of course, alternative strategies for areas still built on 'traditional' industries. The diffusion approach, and the possibility of developing particular integrating sectors, have been mentioned in the discussion here. Such approaches may well hold out more hope for successful intervention in the local economy, but as chapter 3 argued the job of diffusion does not demand a science park (there is no need to be near an institute of higher education).

The wider point, however, is that there are numbers of possibilities open for the precise form science parks can take, only a very few of which have yet been explored. Chief among the questions to be answered by anyone planning a science park (unless they are planning it simply for commercial gain) must be: what will its relation be to the local economy? In other words, how to relate the structures we have explored so far, and which are integral to the very definition of a science park, to an actual area's existing economy? And answering that question will in turn require an analysis of the local economy – its sectoral history and its place in the currently shifting structures of uneven development.

So science parks can play a whole variety of roles in these wider developments, and the role of any one of them will depend on the nature of local economic and social history, on the local area's place in the evolving form of uneven development, and on political strategy. Yet all of them embody contradictions. By their very nature science parks represent an attempt at a spatial separation of different elements of the technical division of labour. Most importantly, R&D is to be geographically separated from production. In many cases this spatial separation reinforces the social divide through an explicitly élitist presentation. But even when it is not so, the geographical separation of groups (R&D and direct production workers) can only serve to reinforce the social polarisation which is already so pronounced within the occupational structure of high-tech industries. Aston and Cambridge Science Parks are in many ways polar opposites, particularly in terms of their regional location, yet the analysis in each case points up this central social issue. Chapters 3 and 4 and 5 have already explored in some detail the nature of the emerging technological élite to which science parks in their classic guise so clearly pander. This chapter has pointed to some of its local effects and questioned, in a number of very different local situations, the ability of the classic science-park model to do anything at all towards solving the problems of unemployment for those not within this élite and of polarisation between the two groups.

NOTES

1 Much the same could be said of the North West of England.
2 In 1983 Birmingham City Council set aside £1.5 million to spend on publicity to change the image of the city (Spencer 1986).
3 The university was still at this time automatically represented on the city council.
4 In fact the area had been a tank yard in the second world war and the real expense of rehabilitation would have been to restore it to agricultural use.
5 This is not necessarily the case. Some parks have adopted different strategies. It should be borne in mind that the Cambridge park has a longer history than Aston.

6 The data used for the detailed comparisons in this chapter are the UKSPA survey (i.e. UKSPA-OU-CURDS) and our own special detailed interviews with firms on both Aston and Cambridge .

7 For definitions see Massey (1984, 1988).

8 The most recent survey on Aston Science Park found that in forty-six companies surveyed 8.6 per cent of markets were overseas and 9.7 per cent of suppliers (Birmingham City Council 1990).

9 This conclusion highlights the difficulties of the (usual) use of surrogates for levels of technological sophistication. There is actually a higher proportion of QSEs on Aston Science Park, but this is due to the different size of the establishments and the difference in the balance of their activities, not to technological level (see chapter 3).

10 Of course, rents on almost all science parks are higher than in their general areas, but in Cambridge the comparison is with an already high base.

11 Indeed, Invest in Britain Bureau (1986) explicitly markets science parks as locations for inward foreign investment.

12 In the 1990 survey of forty-six establishments it was clear that the proportion of women employed on the park had been increasing and in these units had reached 31 per cent of total employees. However, it was impossible to cross-tabulate these by occupation or qualification, and 27 per cent of them were part-time workers (Birmingham City Council 1990). The same survey found that 6.5 per cent of total jobs were held by members of ethnic minorities, that 12.6 per cent had been unemployed prior to their employment on the park, and that 11.3 per cent were temporary workers.

13 The argument here assumes that 'revitalisation' will require firms to remain 'commercially viable' on an individual basis. This is not, of course, strictly necessary.

14 There were other, smaller, elements in its strategy, including the establishment of a Technology Transfer Division in 1982, the aim of which is to 'introduce new product lines, with expanding markets, into existing Scottish companies' (SDA 1984: 38). Here too, however, the strategy looked abroad, seeking 'joint venture licensing agreements with overseas companies who wish to serve the UK and European Market from a production base in Britain'.

7

SCIENCE PARKS AND THE PUBLIC PURSE

INDIVIDUALISM AND ENTREPRENEURSHIP

The model of scientific discovery and innovation which underlies the science-park project in the UK is, as chapter 3 has shown, one in which individual genius figures largely. The same is true, as we saw in chapter 5, of the economic aspects of the project. Much of the imagery and rhetoric of science parks is constructed around notions of individual entrepreneurship. Lord Young as Secretary of State for Employment exulted thus: 'Science Parks have much to do with the wealth and job creation that comes [*sic*] from enterprise, small firms and new technology'.[1] There is a constantly recurring image of bright young entrepreneurs inventing things in garages and winning out to become world leaders. The hype of Silicon Valley in Larsen and Rogers (1984) is about individual success stories; indeed, those authors explicitly point to the utter contrast of this scientific-entrepreneurial culture with William H. Whyte's *The Organisation Man,* published in 1956. Much the same rhetoric has developed in the UK. 'Cambridge is Britain's first emerging Silicon Valley. Between 1970 and 1980, 41 high-technology firms started, many by brilliant academics' (Levi 1985). 'It is generally agreed that Britain could do with more people like Eastwell, an amiable figure who pilots his own aircraft to business meetings and relaxes by driving his red Porsche (fast) around the country lanes of Sussex' (Marsh 1986c). And so on.

The nature of the labour market in high tech reinforces this imagery and culture. As we saw in chapter 4, it is highly competitive, and highly individualistic. It is based on what you know, who you know, where you last worked; people are head-hunted. This is the opposite of undifferentiated labour power. Within the organisation of the companies, with all their autonomy and responsibility, this participation is very competitive and hierarchical. People have to be 'achievers' in the competitive sense to survive. Sociological analyses which have been done of 'high-tech' regions in the US (for example, Dorfman 1983 on Massachusetts) point to the emergence of similar cultures centred on the celebration of competitive individualism. The anti-trade unionism documented in chapter 4 is part of the same self-characterisation, and chapter 5 has shown the importance of élitism in the culture of this class stratum.

The fact that the reality by no means always lives up to the image has been documented in chapters 2, 3 and 5, where the role for instance of big capital in the high-tech phenomenon has been pointed to. And the potential contradictions between research and the individual enquiring spirit on the one hand and the necessities of business and the discipline of the market on the other are illustrated in numerous stories of individual companies. There can be conflict also between the free flow of intellectual enquiry and the business requirements of patents, and between the enterprising spirit which sets up a small company and the tedium of larger businesses. 'The CAD companies were run by clever, individualistic people who didn't like losing their freedom. Life in Cambridge running these companies was quite agreeable. Who wanted to be involved in all the hassles of mergers?' (David Thomson, ex-chair of Compeda, quoted in Marsh 1986b) or: 'Says Richard King – "What appears to be going on is a number of very bright people who have created £3m or £4m companies, who look out of the window of the nice house and see the red Porsche and think – that's a nice size to be. Why should I build a £50m or a £100m company, with all the hassle?"' (Lloyd 1986). But whatever the internal contradictions (freedom/market discipline; individual entrepreneurship/big business) the imagery lives on.

Indeed, it is frequently asserted to be almost a characteristic of places – some places, in this wave of new innovation and investment – and not of others: 'economic success lies with the country and the region and the city that innovate, that keep one step ahead of the action' (Hall 1985: 5). In fact, as earlier chapters have mentioned, a high proportion of the scientists and engineers in the Cambridge area, and indeed in the wider sunbelt region, are in-migrants into the place. And the fact that they can go there and in particular can work on science parks is in part due to that very development of the division of labour which makes their geographical separation from direct production, if not – as we have seen – advantageous to production, at least possible.

The other important element of the prevailing high-tech ideology in the UK (and the US) is that of competition. The competitive free market is the best mechanism of growth; state subsidies and public expenditure only preserve the outworn and weigh down on the finances of the potentially flourishing. Featherbedding is out. Once again, an important reference point of the imagery is the US, and once again it is in part a misrepresentation, or at least out of date. For there have been tensions even in the US. In early 1987, when US chip-makers were losing out badly in the international competition, they turned for help both to each other and to the state. 'US groups unite on chips threat', 'Chip makers ask Congress for help' and 'A U-turn in Silicon Valley' ran the headlines (all these from the *Financial Times,* 5 March, 9 February 1987). 'A few years ago such industry-wide co-operation [Sematech] would have been unthinkable, let alone the idea of government intervention to shore up the proudly independent US chip-makers.' 'Co-operation has replaced competition as the by-word for survival in the US semiconductor industry.' 'It is just three years since executives ... told congressmen that the Government should maintain an "arms length" involvement in the industry if it was not to stifle creativity. Now it seems inevitable that the industry is turning to government for funding' (Louise Kehoe, interviewing Charlie Sporck, president of National Semiconductor and a founder of Sematech, *Financial Times,* 9 February 1987). The debate, and the fate of Sematech, have fluctuated ever since. But the idea that this could have been a complete volte-face itself fails to remember the important role which the US Department of Defense and NASA have played in the genesis and growth of the US high-tech electronics industry.

Yet, once again, the imagery lives on. Moreover, as in the case of individualism and innovation, the ability to stand on one's own feet, to flourish without state help, is deemed to have a definite geography. While the regions of 'the north', perhaps Scotland and the north east of England most especially, are frequently exhorted to become less 'dependent on the state', the economic growth in the outer south and east is hailed as a flowering of free enterprise. The ideology of science parks, too, extols this image. In 1990, at the fifth annual conference of UKSPA, Harry Noble, Director of Economic Development and Planning in Coventry, reflected that in the Secretary of State's address to the 1985 conference the potential role of local authorities went completely unnoticed.

The problem with this element of the ideology is that it isn't true. Throughout the whole of the science-park enterprise there is a high degree of dependence on the state. There is public subsidy of the physical development of parks, public financing of R&D, there are government grants and loans, public-sector contracts, public-sector financing of start-ups, privileged access to publicly funded research ... both the list and the amounts involved are considerable. The notion of the daring, risk-taking, competitive entrepreneur scornful of state aid and expenditure is extraordinarily easy to undermine by the simple recitation of a few facts.

Our argument in this chapter, however, is more complex. It is that the form taken by this connection with the state varies quite systematically between different parts of the country. Indeed, we shall argue, there are clear spatial differences in the articulation together of the public sector and the private. Science parks are new arenas, or attempts to constitute new arenas, for the accumulation of private capital. But the way in which they function in this regard, and the way in which the process is underpinned by the state, varies between different parts of the country. Essentially, the spatial variation coincides with the geography of the new round of investment, the new era of uneven development: there are clear contrasts between the areas which are part of the upswing, and those which are not. Yet both are dependent on the state.

PUBLIC FUNDS AND PRIVATE ACCUMULATION

Science parks and public funding

By 1990, 59 per cent of the accumulated investment in science-park infrastructure and building had been made by the public sector. The 'public sector' included a variety of sources – local authorities, Enterprise Boards, the EC, academic institutions, and quangos such as the SDA, WDA and English Estates – but none the less its contribution has been a very significant one. Probing the data further, however, reveals substantial differences. Although there are exceptions, in broad terms there is a clear contrast between parks which are located in those parts of the country currently experiencing a wider process of high-technology growth and those outside this area. It is the former group which has higher levels of private-sector financing and the latter which is more heavily supported by public funds. Henneberry's survey data from 1982, which covered both science parks and other high-technology developments, show a similar pattern (see table 7.1). There is a quite extraordinary degree of concentration of private-sector investment into the South East and the South West regions. These two regions are also the only ones in the whole country where the location quotients for the private sector are higher than for the public, and they are also the two regions with the highest overall location quotients, where the quotients themselves are calculated against a base which itself reflects the existing distribution of a set of industrial sectors which include the top four high-technology industries in manufacturing. Private-sector investment, in other words, in this wider range of high-technology developments, looks set to increase an already existing geographical concentration. The public-sector location quotients, it will be noticed, are far more evenly spread.

In the most general terms, it is evident why all this should be so. The private sector has moved into the areas which are growing in the currently shifting pattern of uneven development. It is here that profits can be made in development. In other areas

Table 7.1 Public and private-sector funding for science parks and high-tech developments: geographical contrasts

	Location quotient		
Region	*All*	*Public*	*Private*
Northern	0.54	1.07	0
Yorks. and Humberside	0.68	1.10	0.27
North West	1.33	1.83	0.92
East Midlands	0.43	0.87	0
West Midlands	0.24	0.47	0
East Anglia	1.15	1.54	0.77
South East	1.41	0.72	2.07
South West	1.76	0.81	2.70
Wales	0.83	1.67	0
Scotland	1.08	1.89	0.54
Great Britain	1	1	1

Note: The figures include high-tech developments as well as science parks. The location quotients were calculated by dividing a region's share of developments by its share of employees in Industrial Orders VII–XII. A score greater than 1 indicates 'over-representation' of developments, less than 1 'under-representation'
Source: Henneberry (1984b: table 2)

Figure 7.1 Industrial rents: the south-east 'premium'

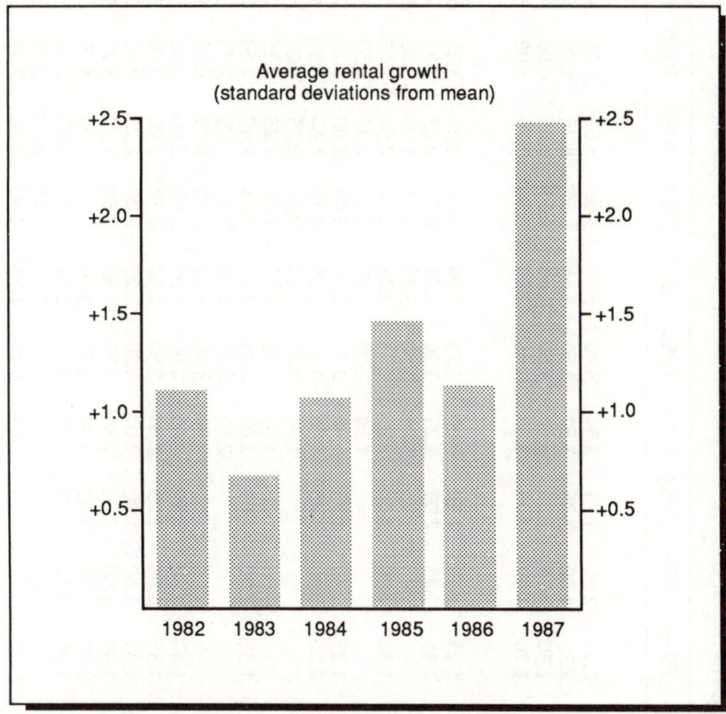

Source: based on Debenham Tewson & Chinnocks (1987: 6)

public-sector funding is needed precisely because of the paucity of profitable opportunities for the private sector. There are a number of reasons for this disparity in opportunities. The first and most obvious is that it is in the former areas, the south and east of England, that there is the largest demand for property for high-technology companies, broadly defined. But it is also in the (outer) south east that supply most lags behind demand, and this in part precisely because this outer metropolitan area is a new region for industrial development. In contrast, in regions where there have been previous rounds of industrial development, there is *both* less demand (in part precisely because of the previous manufacturing history) *and* (for the same reason) more supply already in existence. Of course much of this currently existing supply is unsuitable for science-park-type companies, for reasons

Table 7.2 Prime industrial rents: the contrast between London and the rest of the country (£ per sq ft)

	1973	*1974*	*1975*	*1976*	*1977*	*1978*	*1979*	*1980*	*1981*	*1982*	*1983*	*1984*	*1985*	*1986*	*1987*
London															
Croydon	1.20	1.30	1.50	1.80	2.00	2.15	2.40	3.25	3.50	3.40	3.50	3.60	3.60	3.75	4.25
Heathrow	1.25	1.50	1.75	1.75	2.00	2.25	3.00	3.25	3.50	3.50	3.65	3.85	4.25	4.75	5.25
Park Royal	1.20	1.50	1.70	1.70	1.90	2.30	2.75	3.25	3.50	3.50	3.65	3.65	3.85	4.00	4.50
Tower Hamlets	1.00	1.10	1.30	1.50	1.60	1.70	1.90	2.25	2.40	2.45	2.45	2.60	2.75	3.00	3.75
Provinces															
Basingstoke	0.80	1.10	1.00	1.15	1.20	1.45	1.95	2.70	3.00	3.25	3.00	3.00	3.00	3.25	3.50
Birmingham	0.70	0.90	0.95	1.10	1.25	1.45	1.75	2.00	2.25	2.20	2.25	2.25	2.35	2.50	2.75
Bracknell									3.00	3.50	3.75	4.25	5.00	5.00	5.75
Bristol	0.70	1.00	1.15	1.30	1.25	1.30	1.70	2.00	2.25	2.25	2.25	2.40	2.50	2.65	3.50
Cambridge									1.80	2.00	2.00	2.10	2.25	2.50	3.25
Cardiff	0.70	0.80	1.00	1.00	1.20	1.25	1.50	1.75	1.90	2.00	2.00	2.00	2.00	2.25	2.50
Crawley	1.15	1.25	1.40	1.50	1.65	2.10	2.60	3.25	3.40	3.25	3.25	3.25	3.30	3.50	4.75
Eastleigh									1.90	2.25	2.40	2.55	2.75	3.00	3.50
Edinburgh									1.65	1.75	1.75	2.00	2.25	2.25	2.45
Glasgow	0.65	0.80	0.95	1.00	1.15	1.35	1.50	1.65	2.00	2.25	2.25	2.35	2.40	2.50	2.75
Harlow									2.50	2.60	2.60	2.80	3.00	2.75	3.25
High Wycombe									2.50	3.00	3.25	3.50	3.75	4.15	4.75
Leeds	0.75	0.85	1.00	1.20	1.25	1.40	1.80	1.90	2.00	2.00	2.00	2.00	2.00	2.20	2.40
Luton	0.90	1.00	1.25	1.35	1.35	1.35	1.85	2.00	2.30	2.50	2.50	2.70	2.90	2.90	3.00
Manchester	0.70	0.80	0.80	1.10	1.20	1.35	1.60	1.90	2.00	2.20	2.20	2.20	2.25	2.40	2.75
Milton Keynes	0.75	0.95	1.10	1.15	1.25	1.40	1.80	2.00	2.30	2.40	2.40	2.50	2.75	3.25	3.50
Reading	1.00	1.40	1.50	1.50	1.75	2.00	2.50	3.00	3.50	3.50	3.50	3.50	4.50	4.70	5.50
Southampton	0.80	0.90	1.05	1.20	1.25	1.30	1.40	2.25	2.40	2.40	2.50	2.65	2.85	2.85	3.50
Swindon									2.05	2.10	2.15	2.20	2.40	2.75	3.50
Warrington									1.60	1.75	1.75	2.00	2.25	2.25	2.30

Source: Debenham Tewson & Chinnocks (1987: 7)

both of practical function and of image. Finally, the costs of bringing back into use land which now suffers from the ravages of previous rounds of industrial investment can be considerable. (Chapter 6 gave the example of the need to dig up the foundations of old rounds of investment in the middle of Birmingham.) All this means two things: first, the different dynamics in the relation between supply and demand mean much higher and faster-rising rental levels in the south and east; second, construction costs can be higher in areas of previously intensive industrial development. Figure 7.1 and table 7.2 give some idea of the enormous differences in rental levels and rental growth between the south east and the rest of the country (although it should be noted that figure 7.1 refers to all industrial property). Debenham Tewson & Chinnocks (1987) report that, through the 1980s, regional distinctions in market performance of industrial rental levels have varied in direct relation to the performance nationally of the industrial (manufacturing) sector as a whole. The better the rate of growth of the industrial sector as a whole, the greater the 'gulf in rental performance between the south-east and other regions' (p. 6). In the year to 1987, rental rises in Cambridge were 30 per cent, Southampton 23 per cent and Swindon 27 per cent in an overall regional average of almost 19 per cent. In contrast, the rise in Manchester was 15 per cent, in Leeds 9 per cent, in Birmingham 10 per cent, in Cardiff 11 per cent, in Edinburgh 10 per cent and in Glasgow 10 per cent.

Such calculations are vital to the science-park movement for, while the ostensible reason for their existence concerns the generation of new, innovative production, the key to that process, in the classic model, lies in *property development*.[2] This is the third element, along with their linear model of innovation and their spatial form and content, in our conceptualisation of science parks. Science parks are, by definition, property initiatives. This perhaps seems at first sight like a statement of the obvious, and indeed at this level it is an element of our definition which is shared with the popular conceptualisation.[3] What we hope to demonstrate in this section, however, are the implications of that fundamental characteristic, and its centrality. Indeed, much of the discussion and concern about science parks on behalf of their private-sector financial participants is centred far more on a logic of accumulation through real estate than any concerns about production.

In the south east, science parks are one among a number of new types of development (technology parks and business parks are others) which have provided highly attractive arenas for investment in recent years. Indeed, it has been argued that the land market has become a major element in itself of the overall dynamic of accumulation in the south east (Ward 1987; Thrift 1987), and science parks are high-quality areas in an overall context of regional growth. The financial and property sectors, among others, have not been slow to seize the opportunity. In our discussions about Cambridge Science Park one of the few reservations voiced by tenants was that sites on the park were only available for rent. There is probably more money to be made by land ownership than through production. Trinity College will not sell. And in another example from the sunbelt, the public-sector contribution to the financing of Surrey Research Park was the sale of the land by the university to its two partners, Grand Met and BOC.

But such attractiveness to the private sector is confined to a few parks only. 'Science Parks and Innovation Centres are regarded as a property investment, says Mr Eul, fellow of the Royal Institute of Chartered Surveyors, but they are not attracting funds from property-investing institutions. Except in the property-scarce South East, science parks have been excluded from investment fund consideration, he adds' (Morgan 1986). There are a number of reasons for this conservativeness. There is the fact that these are new kinds of development and assessing their earning potential can therefore be difficult. Rowe (1985), who makes this point, continues, 'Understandably, institutions have started to invest where they can at least see other rents in the same area performing well *i.e.* in the S.E. and particularly along the M4 corridor' (p. 6). There is also the fact that institutions have had a tendency to view property as a medium of investment primarily as a hedge against inflation, the prime requirement being long-term stability. Moreover, during the early to mid-1980s institutions restructured their investment portfolios; there was a movement away from the central allocation of new investment funds to property and towards investment committees of insurance and life funds investing in overseas equities and British securities (Pryke 1989). Finally, the newness of science parks may work against them.

But the more general, and most fundamental, reason why the

private sector as a whole has confined its investment in science parks to a limited and highly defined part of the country is that outside that region science parks are at best risky and at worst unprofitable. There are specific reasons for this. Using a science park as a medium of investment is not just a question of acquiring land and buildings and renting them out. The whole point of science parks is that they must to some degree be 'high-quality' developments, in terms of infrastructure, building construction and possibly also services. Image and status, in other words, again enter the argument, this time as a causal factor in economic calculations. There are, in other words, quite high initial and continuing costs. Moreover such costs will not vary greatly between different parts of the country. In a paper to the 1985 UKSPA conference, Rowe calculated 'a minimum development cost in the region of £40–50 per sq ft excluding any consideration of land and infrastructure' (Rowe 1985: 8). The problem, of course, is that while such costs can be absorbed in regions where land values and rents are high, in other regions they pose a major problem. Rowe continues:

> Asking yields from institutions can vary considerably from 7½% to 10% for Science Park type property depending on an investor's view of potential rental growth for the particular development. As is usual, the parts of the country where the yields are lowest usually command the greatest rent and for these locations the minimum specification poses no problems whatsoever. Indeed, because enhancing the specification does little to produce a significantly more desirable property for the Science Park user, it does not enhance rental values by much and therefore profit and high land values can be realised. Surrey Research Park and Cambridge Science Park must be prime examples of this phenomenon.
>
> However, for the majority of UK Science Parks, ... which lie away from the trendy M4 corridor and Heathrow triangle, institutional investment yields, if offered at all are more likely to fall in the 8–10% region. This produces a minimum rental requirement of around £4.50 per sq ft if the land is given away ... After allowing for servicing the land and land values more typically found away from the South East, minimum rents are likely to be around £5.00 per sq ft.
>
> (p. 8)

Figure 7.2 High-technology rental values, September 1985

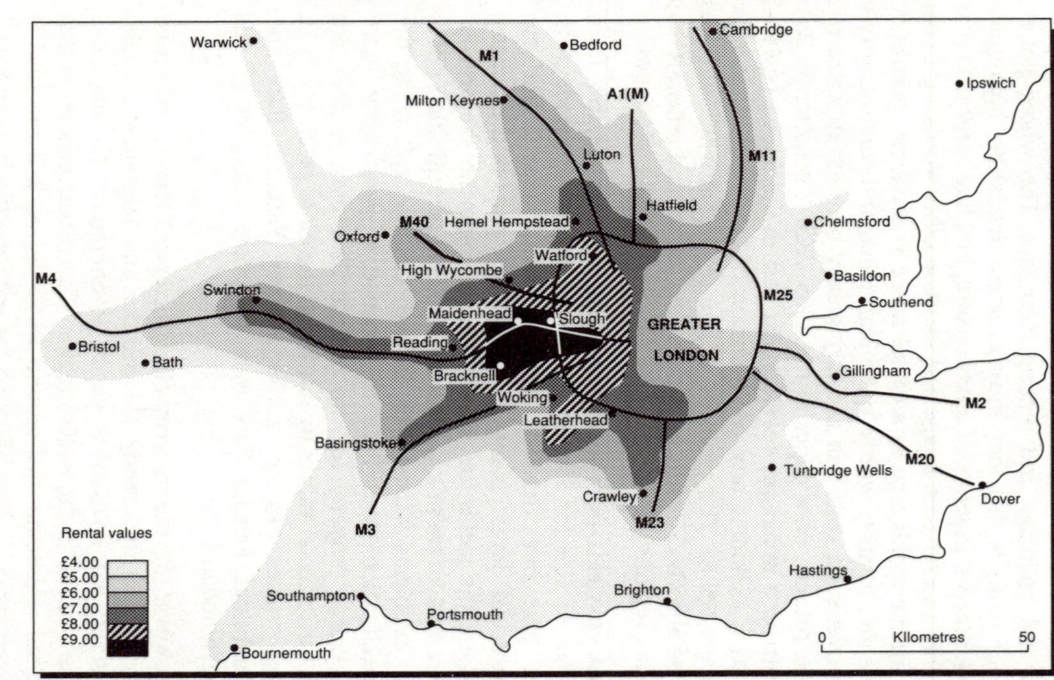

Source: Fuller Peiser (1985)

These figures are for 1985. Figure 7.2 gives an indication of the contours of the high-technology rental market in that year in the south east. Even within that region its boundaries are quite constrained.[4] Moreover, the evidence cited earlier from Debenham Tewson & Chinnocks points out that the gap between the south east and the rest of the country widened still further after that date. Within that favoured part of the south east, then, science parks are a highly attractive property investment for private capital where, given that the construction costs can be covered, 'residual land values can rise rapidly where more substantial rents can be realised' (Rowe 1985: 8). And the situation has not changed in more recent years. In an astonishing presentation to the UKSPA annual conference in 1990, Christopher Edwards, Director of Property Portfolio Management at the Prudential, urged science-park managements to make themselves more attractive to the private sector (for in the end it is profit which matters – 'it's money that makes the world go round') but also said that 'north of Birmingham at this time you could not make the thing viable without subsidy of some sort'. Left to its own devices, private-sector responsibility for the science-park movement could only reinforce the divide between the south east and the rest of the country.

But the private sector has not, of course, been left to its own devices. As was seen earlier, the public sector has to date made the major contribution to the science-park movement. There is very considerable variety. In Wales and Scotland the Welsh and Scottish Development Agencies are important, in northern England English Estates. Local authorities have played a major role, notably in non-regional-policy areas outside the south east, often drawing in other funding (for instance, ERDF money from Brussels) or allocating Inner City or Urban Programme money to this purpose. The university contribution is, perhaps surprisingly, quite concentrated. In many cases funding has been put together from a mixture of sources. And there is a discernible geography to it all.

Apart from the obvious spatial concentration of the national/regional quangos, mentioned above, it seems very broadly that the university contribution (except for the case of Heriot-Watt) is more in the south and east, local authorities are important everywhere outside the south east, and the public–private mixture is most common in the south east and the midlands. Some of this broad geography is indicated in table 7.3. It is not an

Table 7.3 Concentration of infrastructural investment in parks

	*'Southern' parks**		*Parks in Northern Ireland, Scotland and Wales+*	
Category	*Investment (£ million)*	*% of total investment in category*	*Investment (£ million)*	*% of total investment in category*
University	62.9	76	6.2	9
Private	14.5	49	1.6	5
Tenant	3.9	9	6.5	73
Local authority	0.8	3	29.6	25
Government	0	0	0	45
Total	88.4		75.6	

*Ten parks, 26 per cent of total. + Twelve parks, 31 per cent of total
Source: UKSPA

unexpected geography, given what was said above about the potential returns on science-park development in different parts of the country. But what it also represents is different coalitions of interests, the fact that there are different fundamental rationales and objectives behind the existence of science parks in different parts of the country. At the broadest level, the dominance and concentration of the public sector in regions outside the favoured southern area clearly reflect the attempt by that sector in one way or another to compensate for the unequal geography of private-sector investment, in other words to compensate for uneven development. Moreover, the development of science parks coincided with the more general flowering of economic development initiatives among local authorities. In the south east as we have seen the dominance of the private sector reflects the profitability there of science parks as real estate. The greater presence of university finance in the south may reflect the same (or anticipation of the same) phenomenon for in this capacity the universities, though part of the public sector in terms of their financing, are under pressure, both financial and political, to become more 'commercial'. Their objectives in science-park investment may therefore be complex (they may, for instance, include the aim of becoming more high-profile in science and technology, or of establishing more links with industry in general) but they are likely to include a positive financial return on investment. This was, as we saw in chapters 2 and 6, the case with Trinity College, Cambridge. Such investment, of course, depends on having the capital there in the first place. And here the universities are very unequal in their endowments – we saw the contrast between Trinity and Aston in the last chapter. They are also of course very different, by virtue of their locations, in the potential financial gain which can be made from any land they do own. Universities in the south and east, and which own land, can benefit financially from initiatives such as science parks in a way which northern universities simply cannot. Movements in land values are thus a criterion in the distribution of research funding. Farther north again, the public sector is more dominant and its role is different. The WDA, SDA, English Estates and local authorities are backing developments which the private sector refuses to handle. Here the relation between public and private, and between property development and production, can take a number of forms. It may be that the costs of providing a science-park-quality development are such that an investment in such a real-estate deal can never in the foreseeable future be profitable. In this case the public sector is subsidising production (the firms which come to locate on the park) by providing premises at rents which make them uneconomic from a property point of view. It is a classic role for the public sector and reflects the objectives of such science parks as part of socially orientated local economic strategies. A second form of public–private relation is a variation on this. It may be that the public sector steps in where the private sector fears to tread, not because the development is in fact irretrievably unprofitable but because the private sector will not respond to its potential profitability. This may simply be because positive profits are in themselves not enough to attract private investment – the level of positive profits has to be competitive with that elsewhere – or it may be because the conservativeness of the private sector, mentioned above, prevents it taking either the imaginative leap or the risk which may be involved in science-park developments in the north. In both these cases the public sector is taking up potentially viable commercial investments which are simply ignored by the private sector. The 'subsidy' which is involved here is simply that of opportunity costs potentially forgone. Given the political constraints on local authorities, these are unlikely to be large where they are the financing institutions. What the public sector is doing is providing a commercially viable product which, because of the combination of the nature of the investment (the newness and perceived riskiness of science parks) and its geographical location, the private sector is failing to provide. There is a third relation between public and private which takes this logic further. This is where the public sector turns the science park into a development which the private sector *will* invest in. In this case the public sector is effectively subsidising private capital to make profits out of property development, so that sites can be provided for production.

Such ventures are obviously most feasible in areas on the margins of profitability, and indeed as was mentioned above a public–private mix in the financing of science parks is most common in the south and the midlands. The role of the public sector here is to eliminate the risk and to create local prime sites for the private investor – in other words to make science parks into islands of investment potential which can vie in attractiveness with property investments in the south east.

Warwick Science Park presents an excellent example of this phenomenon, and the process of its funding has been well documented by Rowe (1985). The University of Warwick Science Park was initiated through the establishment of a company by two elements of the public sector – the university and the local authorities in the region. The West Midlands County Council and Warwickshire County Council funded the first tranche of investment and Barclays Bank invested in the first building – an incubator centre of very small units. This incubator unit is described by Rowe as 'a true full risk property investment' (p. 3), which it was in terms of the form of investment, though the infrastructure for it had of course been provided free by the local authorities. Given that subsidy, it has been successful in investment terms and indeed has been extended.

The progress of the park since then has continued this articulation of public and private. In this 'partnership' there are two principal roles played by the public sector. The first is that it is the public sector which provides the risk capital. The private investors only come in when an acceptable rate of return has been virtually guaranteed. Thus it was decided to extend the incubator idea to the provision of rather larger premises. However, 'it was recognised that any decision to move into the construction of speculative units of 5000 sq ft or more was a high risk proposition and private capital was not ... considered' (Rowe 1985: 3). 'However, the local authority shareholders in the venture could see the strength of the logic and were also convinced that if the Science

Park did not make progress rapidly at the beginning then the whole initiative could fade from lack of momentum. It was the West Midlands County Council who stepped in first with a further £1m for a block of larger units from 5–10,000 square feet' (p. 3). Within six months from completion the building was fully tenanted.

However, 'the ever tightening squeeze on the availability of local authority capital funds and university finance meant that these sources could not be relied upon indefinitely' (p. 4). It would be necessary to get funds from the private sector. A central element in this was an attempt to get funds from the financial institutions. It was a slow process. And once again the public sector stepped in, the result of a combination of its greater commitment to the project and its greater imagination in seeing the investment potential. In contrast, the financial institutions were interested in the project only as a commercial venture, and even within those parameters were operating in an extremely conservative fashion. Once again it was the public sector which took any risk there was, at the same time thereby further reducing any risk attached to subsequent investments:

> The fact that it has taken the Science Park two years to make any real progress with the market has meant that in the interim period Coventry City Council decided to go ahead and invest a further £1.2million in another group of speculative units ... That building, which at the time of writing in December 1985 is only two thirds constructed, is already fully committed with the space being taken up by new major companies coming to the Science Park (60%) or companies who have outgrown the units in the Barclays Venture Centre (40%).
>
> (p. 4)

In other words, the role of the public sector was to be the risk-taker, demonstrating in practice the commercial attractiveness of the science-park property. At the end of his paper Rowe gives advice to agencies wishing to approach the financial institutions for science-park funding. He begins:

> First, for other Science Parks like Warwick whose University falls outside the high rental areas of the country, they must believe in what they have to offer and recognise that the property component of a Science Park has a market value

> which may be higher than dictated by local conventional property wisdom. However, before approaching the institutional market for funds they should at least have some evidence to support the view that they can realise and maintain economic viability.
>
> (Rowe 1985: 9)

But who then is the visionary, risk-taking entrepreneur? It seems it is the public sector, not the private investor.

It is not, however, merely current commercial viability that the financial institutions and other private-sector potential investors are seeking. They also need to be assured of future rental growth. Perhaps especially in normally marginal regions in terms of industrial property investment, this means constructing science parks as local 'prime sites'. This involves both quality buildings and all the signs and symbols of prestige. Even the property component of the concept of a science park is in this sense relational – its characteristic is superior status in relation to other sites in the local area.

Similarly, while rents on Aston Science Park were initially set in relation to the surrounding area, the opposite is now the case (Albert Bore, Chair of the Economic Development Committee, Birmingham City Council, speaking at the 1990 annual conference of UKSPA). Thus, to turn again to Warwick:

> It was … clear that the economics of investment in property on the Science Park was marginal. Coventry, the nearest city, has weak city centre rental levels for a major city in the country and the industrial rents are also modest. Furthermore, many major institutional funds who had acquired West Midlands property in the boom years of the 1960s were now divesting as property performance in this part of the world showed poor returns. To set against that, the Science Park was experiencing a high level of demand and rents were starting to reach an encouraging level.
>
> (Rowe 1985: 4–5)

To overcome the generally dismal regional context the science park had to be constructed as a high-quality 'prime property i.e. it should be in the right location for its use, constructed to a high standard and with a sound occupational leasing structure under which all obligations are passed to the tenant leaving the landlord to enjoy a "net rent" with reviews that can only move upwards'!

(p. 5). Such quotations not only leave in tatters any notion that private-sector property investors are risk-takers; they also show how the requirements of these institutional investors can in turn put pressure on the other activity which science parks are supposed really to be about – their tenant firms.

On this scenario what this type of science-park development comes down to is the provision of new, public-sector-created arenas for accumulation through property investment; protected havens within older areas of industrial development. All the evidence is that science parks everywhere have higher rental values than their surrounding regions (see, for instance, Henneberry 1986: 5). Thus, for example, 'Bradford's developers have been pleasantly surprised to find their science park has actually improved the tone of the neighbourhood, their initially high rents gradually driving up nearby property returns' (Morgan 1986). In these surrounding areas rents are presumably rising by image association rather than as a result of the provision of the high-quality infrastructure and buildings which are part of the reason for high rents on the park. It is a 'free' rent rise to the property owners, but it is also an increased cost on production.

To guarantee still further the flow of future rental income, Rowe (1985) also recommends the early construction of incubator units:

> There is now ample evidence to show that these units literally grow the larger tenants for other premises on the Science Park. In this way, the element of void risk in the construction of speculative units on a Science Park can be significantly reduced ... Therefore, uncertainty over demand is significantly reduced for any institutional investor where an incubator project is already present on the Science Park.
>
> (p. 7)

And finally, in case there should be lingering worries, there is that relational characteristic which is essential to the definition of science parks – their exclusivity. 'It will not be many years before this land is developed and thereafter this resource will become scarce and there will be a growing potential for rents to rise by virtue of this scarcity' (p. 7). 'Ultimately, the restricted availability of University linked Science Park land should provide some measure of comfort that scarcity will *at least* maintain rental levels' (p. 10, emphasis in original).

There have been three distinct but interlocking dimensions which have structured the argument in this section: there is the spatial dimension, the issue of regional disparities; there is the relationship between the public sector and the private; and there is the relationship between science parks as a medium of property investment and science parks as places of production and potential regenerators of their local economies. What we want to argue now is that there are three (at least) classic models, each of which embodies a particular articulation together of these three dimensions. Moreover, to understand the functioning of these distinct models it is necessary to return to the fact that science parks are by definition property initiatives. Quite what that entails, however, will depend on the social form of the property initiative, and this in turn will depend (although not deterministically, as will be seen) on the *agents* which are involved. For it is the nature of the agents, and the place of the initiative within their overall operations, which will in turn influence the *aims* set for science parks as property initiatives. It has already been seen in chapters 2 and 6 how the aims of science parks vary between agencies. Finally, it is important to note that what underlies the constitution of the classic forms we are about to describe (that is, types of science parks as property initiatives) is geography. Specifically, it is uneven development.

First there is what might be called the classic 'sunbelt' model. Here, in the outer south and east particularly, science parks tend both to be private-sector funded and to be property-led. Their objective is that they produce a return as property investments. They feed off, or are part and parcel of, already growing local economies in which both the land market and 'high tech' are central dynamics. They are profitable *as* property initiatives.

The second model is that of the public–private mix, especially where one of the aims of public-sector intervention is to attract private-sector financing of the science park. Almost by definition this model is most likely to be found – as indeed it is – in regions on the margins of profitability in rental terms, as perceived by the private sector. What this model shows is that it is not really the public–private dichotomy which is the issue, but the fundamental objective of the science park. And here the basic division is between science parks as property investments and science parks as regenerators of local economies. Universities, though publicly funded, are likely to prioritise the science park as property investment (in reality if not always in rhetoric) and are effectively pressed to do so by government cuts. The focus of other elements of the public sector – local authorities in particular – is likely to be more on the potential local economic benefits. Trying to pull in private finance, however, creates contradictions. It necessarily pushes the local authority into presenting (and developing) the science park as a property investment. (The private-sector speaker from the world of finance at the 1990 UKSPA conference demonstrated the power relations perfectly as he sternly 'told off' the science-park managers for not presenting their case in sufficiently attractive form to capture the interest of the institutions.) This has two effects. First, it leads to the public sector taking on all the risk before the private sector comes in to take a profit. Effectively this means that the financial institutions and other private-sector property developers are taking the profit off investment undertaken by 'ratepayers'. There is public subsidisation of profits from property development. It might be argued that the answer should be total public-sector funding. But there are problems with this, too, not least that it is the financial squeeze on local authorities which forces them into this less remunerative form of investment of their money. That is, they are forced to bring in private-sector money because of their own shortage of capital, but in order to do that they have to pass the bulk of the return on the investment over to the private sector, while having made all the initiatory moves, and taken all the risk, themselves. Such, of course, is the real meaning of 'leverage.'[5]

But the second issue is that there may well be conflicts between the objectives of science-park-as-property-investment and science-park-as-part-of-local-economic-strategy. Henneberry, who points to these two very different sets of science-park objectives, writes:

> the temptation is strong to evaluate science parks in property terms ... Indeed, the pressure to let schemes quickly to begin to get a return on capital is considerable. The more stringent are the tenant selection policies the longer the floorspace will take to let and, effectively, the more expensive is the development of the science park concerned. Many private sector hi-tech schemes have been forced to accept 'low-tech' occupiers in order to boost rental income (e.g. Tesco distribution warehouse at Aztec West). Developers of some science parks arranged 'fall-back' positions in case space proved difficult to let (e.g. the planning permission on Listerhills, Bradford, reverted to full office use after 2 years in the event of the scheme failing).
>
> (1986: 2–3)

So a property orientation may shift the criteria by which production activities will be accepted on a park. Those activities which can pay the highest rents are not necessarily the ones which will do most to regenerate the local economy.

Further, the need for high rents and secure conditions may impact on the occupiers. Leases 'with all obligations passed to tenants' were mentioned above, indeed were being recommended as a way of attracting institutional property investment. And high rents from high-tech can have negative effects on other production if the impact on land values ripples out to a wider area. An example of this is given in figure 7.3.

The third classic model is that of total public funding with the prime objective of regenerating the local economy. This is the dominant model in 'the north'. Here both the public–private articulation and the property–production relation are very different. The public sector, to counter the private-sector geographical bias towards the south and east, effectively subsidises production on the park through the public provision of property. It is, once again, a classic local economic development policy, particularly in areas of decline. Without public-sector initiative and funding there would be few, if any, science parks outside the sunbelt south. Without doubt, therefore, public-sector investment has shifted the distribution of science parks towards the north (see also Henneberry 1984b). Whether that has had any impact on the distribution of high-technology growth, or on its absolute level, is another question. The *true* success of science parks, in terms of the normal claims made for them (see chapter 2), must be measured in terms of economic regeneration, local or otherwise. The history of property initiatives here is not encouraging. As Segal writes, 'the lesson from Cambridge, and indeed from every other situation that we know of, is that planned provision of property is not itself a sufficient factor (and sometimes not even a necessary factor) in stimulating development of high technology industry' (1985: 566). In the end, the issue concerns not the property market but the revival of production. It is therefore to the public–private interface within production that we now turn.

Public funds and production on the parks

Data from our national survey of science parks indicate that public-sector financing was also a significant element in production (tables 7.4 and 7.5). The most important form which

Figure 7.3 High-tech and other industrial property

The lack of new supply in the M4 towns was highlighted in earlier surveys, but the limited availability of new light industrial/warehouse units is now an increasingly dominant factor over a much wider area in the south-east and is beginning to affect some northern cities. The pattern of scarcity can be linked to the development of 'high-tech' accommodation. In the intensely competitive commercial property markets of Reading, Cambridge, Bracknell and Swindon, for example, new development has been focused on the needs of the expanding business areas such as computers, pharmaceuticals and electronics.

For these activities, where a considerable element of the production process is desk-based, accommodation has had to provide much higher standards of internal fittings, larger office areas, air-control, more car parking and in many instances a high-profile setting within a landscaped, low density site. Rents for these premises are consequently substantially above those of the more basic industrial buildings; in Cambridge, for example, by a factor of two-threefold. Input and building costs are much higher but in the early 1980s, when such buildings were in short supply, the higher profit margin attracted many private and institutional developers.

Higher rents and development profits have filtered through to higher land prices and, in locations perceived to be potentially attractive for 'high-tech' industries, land prices have accelerated over the last few years. The economics of light industrial/warehouse development has become marginal as a result of the rapid rise in land values and the recent growth in rents partly reflects this escalation in land prices. In Cambridge, where the local industrial market has been overwhelmed by science parks and 'high-tech' schemes, land prices have doubled since 1986 to over £500,000 per acre. Around the M25 and in the Thames Valley industrial land prices have jumped by one-third over the last year and now exceed £750,000 per acre in locations where there is a proven demand for business park accommodation.

Outside the south-east the high-tech sector is not as influential, but even in locations where the demand for industrial accommodation is still weak, the competition from other sectors for sites has bid up land prices, albeit at a slower rate.

Source: Debenham Tewson and Chinnocks (1987: 4–5)

this took was straightforward grants from the public sector. Grants from central government public-sector bodies played a significant role, though they were more important for independent firms than for subsidiaries. Public-sector grants from other sources also showed up strongly. There was some equity participation, particularly by local authorities, in science-park independents (though not in subsidiaries), and some loans from public agencies, though both of these were relatively infrequent.

Even when all these are added together, however, the public sector is dwarfed in importance as a significant source of finance to production by the private sector. Table 7.5 gives the aggregate data for Aston and Cambridge. In both start-up and current situations, and for both independents and subsidiaries, between two-thirds and three-quarters of science-park establishments surveyed cited the private sector as their most important source of funding. The differences between categories within the private sector (table 7.4) are fairly expectable. It must be stressed that what these tables show is counts of the most important source of funding by establishment, and counts of other 'mentions'. They therefore say nothing about the amounts of money involved. Such data were not obtainable on a sufficiently systematic basis to be usable. None the less, even with this information, it is possible fairly safely to infer that public funding of production is far less important, as a proportion, than it is of science-park infrastructure and building (in the latter it amounted to 59 per cent of total funding, or 30 per cent if universities are excluded). The implication, for the moment, is that the public sector is more important as a provider of property than as a funder of production. It is a very arm's-length relationship for a 'high-tech' strategy.

These data are from the national survey and are unfortunately not disaggregatable by region. So from this source it is not possible to investigate geographical differences to see if there are any parallels with the spatial variation in the funding of science-park infrastructure and buildings examined in the last section. A start in this direction can, however, be made by exploring the information from our detailed survey. This allows a comparison of Aston and Cambridge. The basic data are presented in table 7.6.

Immediately, it is clear that there are contrasts. Some of these can easily be put down to the difference between the populations of firms on the two parks, and should not be over-interpreted in terms of geographical differences. It will be remembered from chapter 6 that Aston firms are, on average, younger and smaller than Cambridge ones – there is a higher percentage of start-ups and independents. Cambridge has more large establishments and more subsidiaries. These differences could well be the reason behind the greater importance of personal savings at Aston and of parent-firm funding at Cambridge. But there are other differences too. Most noticeably, for Cambridge establishments private finance is both more often mentioned, and more often mentioned as being the most important source, than at Aston. The difference is most startling in the case of current sources of finance, where 90 per cent of Cambridge establishments said the private sector was their most important source, as against 46 per cent of Aston establishments. This is exactly what might be expected. Private-sector investment in production, just like private-sector investment in property, currently favours the 'sunbelt' as against inner-city areas. That is the geography of the current round of investment.

Yet the situation is also more complicated than this. For in three out of four of the columns in table 7.6 Cambridge also has a higher level of mention of *public* funding than Aston. The real difference is in the availability of mixed public–private sources on Aston. This category does not show up at all on Cambridge, and in Aston it is virtually all accounted for by Birmingham Technology Ltd, the funding source put together by Birmingham City Council and Lloyds Bank as part and parcel of the former's local high-technology strategy. Its importance is immediately obvious from the figures, and of course the availability of this type of funding may itself be a further reason why there is such a significant presence of small independents on Aston. Indeed, availability of finance was a significant reason cited by some Aston companies in their choice of site.

What is being reflected here, in other words, is partly the distinct roles of the two science parks within their local areas but also geographical differences in the unsatisfied need for venture capital. Private-sector venture capital emerges as having been more important in Cambridge than Aston (table 7.6), and indeed Cambridge is now legendary as an area where such funding is readily available. In Aston that role is taken by a joint venture as part of a public-sector local economic strategy. That there is a demand is indicated by the figures. Such detailed data, of course, are consistent with all the recent evidence of geographical variation in the availability of venture capital across the UK as a whole (Mason 1987; Fazey 1987). It is clear that there is here another gap, with specifically geographical dimensions, which the private sector, unaided, is failing to fill.

But there are further contrasts between Cambridge and Aston.

Table 7.4 Sources of finance for firms

Source	Science park independents				Science park subsidiaries			
	Start-up		Current		Start-up		Current	
	Mentions	Most important	Mentions	Most important	Mentions	Most important	Mentions	Most important
Personal savings								
Of the chief executive	50 (19.4)	35 (32.1)	36 (11.7)	14 (13.5)	5 (8.6)	4 (10.5)		
Of other full-time directors	39 (15.1)	14 (12.8)	23 (7.5)	7 (6.7)	5 (8.6)		2 (3.0)	
Of part-time directors	10 (3.9)	2 (1.8)	6 (2.0)					
Of other members of the workforce	2 (0.8)		4 (1.3)			1 (2.6)		
Retained profits	5 (1.9)	3 (2.8)	36 (11.7)	26 (25.0)	7 (12.1)	5 (13.1)	13 (19.4)	9 (25.0)
House mortgage	14 (5.4)	2 (1.8)	6 (2.0)					
Loans/gifts from friends/relations	7 (2.7)	3 (2.8)	6 (2.0)					
Loans/overdraft from clearing bank	41 (15.9)	13 (11.9)	49 (16.0)	21 (20.2)	6 (10.3)	2 (5.3)	8 (11.9)	1 (2.8)
Loan from finance company	3 (1.2)		5 (1.6)					
Loan from public agency (NCBE, BSCI)	2 (0.8)	2 (1.8)	6 (2.0)		1 (1.7)		1 (1.5)	
External equity participaton from:								
Venture capital fund	11 (4.3)	3 (2.8)	13 (4.2)	4 (3.8)				
BES fund	1 (0.4)		2 (0.7)					
Private investor	4 (1.6)		8 (2.6)	1 (1.0)				
Existing business/parent	6 (2.3)	5 (4.6)	7 (2.3)	4 (3.8)	13 (22.4)	12 (31.6)	16 (23.9)	16 (44.4)
University	4 (1.6)		4 (1.3)		2 (3.4)	2 (5.3)	2 (3.0)	2 (5.6)
Local authority	2 (0.8)		3 (1.0)	1 (1.0)				
Hire-purchase	6 (2.3)		16 (5.2)				1 (1.5)	
Factoring			3 (1.0)	2 (1.9)				
Leased equipment	11 (4.3)	1 (0.9)	22 (7.2)		4 (6.9)		4 (6.0)	1 (2.8)

Grant from																
Dept of Industry, e.g. SFA, RDG	5	(1.9)			10	(3.3)										
Dept of Industry R&D grant	4	(1.6)	2	(1.8)	8	(2.6)	2	(1.9)					1	(1.5)		
Dept of Industry BIS	2	(0.8)			11	(3.6)	1	(1.0)					7	(10.4)		
SERC	1	(0.4)	1	(0.9)	2	(0.7)	1	(1.0)	1	(1.7)	1	(2.6)	1	(1.5)	1	(2.8)
European source	1	(0.4)			2	(0.7)										
Don't know									1	(1.7)	1	(2.6)				
Work in progress	1	(0.4)														
Shareholders/chairman	2	(0.8)	1	(0.9)												
Various	2	(0.8)	1	(0.9)	1	(0.3)	1	(1.0)	6	(10.3)	5	(13.1)	5	(7.5)	3	(8.3)
Public-sector grants	7	(2.7)	4	(3.7)	5	(1.6)	3	(2.9)	1	(1.7)	1	(2.6)	1	(1.5)	1	(2.6)
Salary subsidy					1	(0.3)										
WDA, W. Yorks E.B.	1	(0.4)	1	(0.9)	1	(0.3)	1	(1.0)								
Private investors	2	(0.8)	1	(0.9)	2	(0.7)	1	(1.0)								
Loan guarantee scheme	4	(1.6)	3	(2.8)	5	(1.6)	3	(2.8)	1	(1.7)			1	(1.5)		
Family support	1	(0.4)			1	(0.3)										
Unpaid work	2	(0.8)														
Loan from UNI, UGC	2	(0.8)	2	(1.8)					2	(3.4)			1	(1.5)		
Customers up front	1	(0.4)	1	(0.9)												
Accumulated tax					1	(0.3)										
BTS, NCB					1	(0.3)										
Sponsorship									1	(1.7)			1	(1.5)		
Public agency, flotation									1	(1.7)			1	(1.5)		
Other	1	(0.4)	9	(8.3)	1	(0.3)	11	(10.6)	1	(1.7)	4	(10.5)	1	(1.5)	2	(5.6)
Total mentions	258	(100)	109	(100)	307	(100)	104	(100)	58	(100)	38	(100)	67	(100)	36	(100)
No. of firms	109		109		101		104		36		38		37		36	
Mentions per firm	2.4		1.0		3.0		1.0		1.6		1.0		1.8		1.0	

Source: UKSPA-OU-CURDS

Overall, public-sector funding is mentioned more often by Cambridge establishments than by Aston ones. This immediately runs counter to the ideology of the private-sector success of the 'sunbelt', scornful of the state, mentioned at the beginning of the chapter. The image is further undermined when the detail of those public sources is examined. The main source which Cambridge has access to far more significantly than does Aston is central government grants. In every category, grants from the DTI, from NEB (as was), from ICOF and from MOD are more important on Cambridge than on Aston.[6] This marks a significant contrast between the two parks in their relation to the public sector through production. While establishments on Aston have access to BTL funding, this is through the local authority and takes the form of loans or equity. In contrast, Cambridge's more important source of public funding, and which its establishments appear to have more access to than those on Aston, comes from central government and takes the form of grants.

Table 7.5 Sources of finance mentioned as being 'most important', Aston and Cambridge (%)

	Start-up			*Current*		
	Public	*Private*	*Mix*	*Public*	*Private*	*Mix*
Aston Science Park establishments	14.3	52.4	33.3	12.5	45.8	41.7
Cambridge Science Park establishments	20.0	80.0	–	10.0	90.0	–

Source: Survey data and UKSPA-OU-CURDS

Moreover this kind of funding through direct financing of the enterprise as such is not the only way in which the public sector relates to production on the parks. Firms may survive mainly by selling to public-sector purchasers, or they may draw on publicly created resources, most particularly – in the case of these companies – research, without any financial transactions being undertaken. The main similarity between the two parks is that public-sector markets are important for a significant number of firms. In quite a number of cases on both parks the public sector accounts for over half the company's market. On both parks too, a number of firms draw significantly on publicly funded research in the university sector (this, of course, is to be expected, since this kind of public–private relationship is precisely the point of the science-park concept – though it should be noted that not all the links are with the university next to which the science park has been established).

The big *difference* between the two parks, however, is in the access which the firms on them have to publicly funded research in government research laboratories. This kind of 'public subsidy', through the major national scientific laboratories, is highly developed on Cambridge, and absent from Aston. On Cambridge we found one company (Agricultural Genetics Company Ltd) with access to AFRC research, another with access to research by the Ministry of Agriculture, a third with a contract to commercialise the research of the MRC, and a unit established to commercialise government-funded Ministry of Defence research. These links are not trivial. Agricultural Genetics was formed to exploit plant-related discoveries made in UK government-funded AFRC institutes (interview). 'The company's unique agreement with the AFRC gives it first option to develop commercially the plant biotechnology discoveries made at the six AFRC institutes that specialise in this area of research' (AGC 1986: 3). Subject to the terms of an agreement, the company has the option to acquire certain rights in agricultural biotechnology at six AFRC institutes. These AFRC laboratories give Agricultural Genetics access to world-class science which is a big attraction to private investors and which means that the company has no need to set up its own expensive R&D facilities. It is, therefore, essentially a technology-transfer company whose business is market evaluation, research funding and research management, and facilitation of production by the establishment of new subsidiary companies, the acquisition of existing companies, licensing arrangements and joint ventures. A company with rights to commercialise Ministry of Agriculture R&D into agricultural diseases also has a DTI grant of £250,000 and BES funding. As in the case of Agricultural Genetics, it was established when its founders saw the opportunity to commercialise publicly sponsored research in government laboratories: 'This R&D base is unique. It would cost millions to do in-house. We could never afford it. It is the basis of our competitive advantage' (interview). Forty per cent of the company's sales are back to the public sector, in the form of the National Health

Table 7.6 Sources of finance for firms, Aston and Cambridge (%)

Sources	*Start-up*				*Current*			
	Mentions		*Most important*		*Mentions*		*Most important*	
	A	*C*	*A*	*C*	*A*	*C*	*A*	*C*
Private								
Personal savings	27	9	19	7	5	3		
Gifts, loans from friends/relations					5			
House mortgage	4							
Loan/overdraft from bank	15	9	5		5	14	4	15
Loan-finance co.	2							
Credit cards	2							
Equity participation:								
Corporate						7		10
Venture capital	2	9	5	7	5	7	8	5
Individual	2	4	5	7				
Parent-firm funding	6	30	14	47	10	24	17	35
Leased equipment	2				2			
Sweat equity	2	4						
Share capital		9		13	2	17		15
Retained/profits	2		5		17	10	17	10
Total	66	74	53	81	51	82	46	90
Public								
Loans	2		5		2		4	
Equity participation (BES)		4		7				
Grants: DTI, NEB, ICOF, MOD	6	22	5	13	12	14	8	10
WMEB	2							
Loan guarantee scheme	2		5			3		
EC source	2							
Total	14	26	15	20	14	17	12	10
Public-private								
BTL loan	8		19		15		21	
BTL equity	10		14		20		21	
Total	18	0	33	0	35	0	42	0

Note: A = Aston, *C* = Cambridge
Source: Survey data and UKSPA-OU-CURDS. Figures may not sum exactly, owing to rounding

Service. A subsequent round of funding has been able to attract five major investors, including pension funds and insurance companies.

These are just two examples. The immediate point is that the sunbelt firms of the Cambridge area seem to have far greater access to this type of public-sector support than do companies in other parts of the country. As Segal Quince (1985) conclude, 'public research expenditure has been a fundamentally important preconditioning factor in Cambridge' (p. 58). This is true both in the sense of the long-term commitment to government-funded research in the Cambridge area and in the specific cases of individual firms. This differential generosity of the central state in favour of R&D and production, especially 'high-tech' production, in the sunbelt south and east, and its impact on economic growth in that region, has been documented elsewhere (see Hall *et al.* 1987 on the M4 corridor, and Morgan 1986 on technology policy and the south east more generally). Laser-Scan was the first firm on Cambridge Science Park, and is often cited as 'an archetypal model for a Science Park firm' (see chapter 3 and Monck *et al.* 1988: 231) – it is a university spin-off, 'high-tech', and has a successful record. Its dependence on state expenditure is total. Its original R&D and product development were done in the Cavendish Laboratories at Cambridge University, its original customers were universities, further product development was funded by the Ministry of Defence, it received technical support from the government-funded CAD Centre, and its customers are virtually 100 per cent public-sector.

The arguments which can immediately be drawn from this are twofold. First, the obvious point that the competitive, free-market imagery of the sunbelt is seriously lacking in substance. But second, and more interestingly, that there seems to be, not so-called 'freedom' from the state in some parts of the country and 'dependence on the state' in others, but a regionally differentiated geography of relations between the private sector and the public.

PUBLIC–PRIVATE RELATIONS AND UNEVEN DEVELOPMENT

The flowering of the science-park phenomenon has taken place, in the main, during a period of significant rearticulation of relations between the public sector and the private. Indeed, science parks themselves, in so far as they embody an attempt to commercialise publicly funded research in universities, are part of that (attempted) rearticulation. That particular aspect of the public–private relation, between companies on science parks and university research, has already been examined (in chapters 2 and 3) and has been shown not to have taken the form predicted by the model of scientific discovery on which science parks are based. However, as we have seen in this chapter, there are other ways in which science parks and the establishments on them are at the intersection of the public and private sectors of the economy.

As this chapter has shown, science parks and their associated companies are currently, in quite different parts of the country, in one way or another dependent on the state. The fantasy of areas of 'success' such as Cambridge being entirely the result of private enterprise is indeed a fantasy. However, the nature of the public–private relation varies across the country. Everywhere, science parks are established as new areas for the accumulation of private capital, but the way in which this is dependent on the state varies between regions. In those areas caught on the upswing of the current phase of investment the science park itself is primarily a property investment, established as such by private-sector finance. Part of the basis of such areas' success is, however, the fact of government investment in research over the long term and of central government grants and rights of access to research to individual companies. In areas of the country at present left behind by the current round of investment, however, the science park is more likely to be related in its aims to production and to economic growth. Here the public sector is more likely to take the lead in financing. Yet it is doing so through property. Effectively, it will provide either a subsidy to production through the provision of property, or a subsidy direct to the property development. Moreover, in these latter areas the public-sector money is more likely to come either from the local state, the EC or one of the regional-policy quangos.

In many ways this is a perverse scenario, and current changes in central–local government relations are likely to make it more so. The attack on local authorities' policies of economic regeneration (e.g. through Enterprise Zones and Urban Development Corporations) could effectively push them more towards intervention through property. Yet even that role will become more difficult to perform effectively, as was seen in the Warwick example, because of local government funding cuts and other ways in which central government is cutting other sources of funding, and the room for manoeuvre, of local authorities. This is particularly the case with funding from the European Community, of which local authorities are prevented from taking full advantage by central government insisting that local authorities cover EC grants out of their capital programmes, for instance, or devising rules which mean that any funding from such sources will be effectively nullified by being deducted from central funding. On the other hand central government plays both roles in relation to science parks. It funds property through the Development Agencies and English Estates and is actively involved in production through research and various forms of grants and contracts.

It has already been mentioned that the overwhelming evidence is that property policies are insufficient to spark off processes of regeneration, certainly on the tiny scale represented by science parks. Rather than stimulating new industrial investment, their effect is more likely to be simply geographical redistribution. Indeed, there may be conflicts between the aims of science parks as property investments and science parks as local economic regenerators. And if proof were needed of the superior results of public investment in research, and direct encouragement of production, then Cambridge (and much of the rest of the sunbelt – see Hall *et al.* 1987) provides it. The evidence from Cambridge points to two conclusions: first, the degree to which private accumulation occurs on the back of public support and, second, the degree to which serious government support for research and production *can* in fact help promote economic growth. Yet it is precisely this kind of activity (though usually with more explicitly social content) by local authorities which is currently being attacked by central government. Indeed, recent cuts by central government, in support to universities and to research, and privatisation of major scientific research institutes, may undermine even its own effectivity.

The further irony of this is the geography of it all. While the big central government subsidies to production go to those areas flourishing in the upswing of the current phase of uneven development, out in the rest of the country it is local authorities and regional bodies which have to try to offset the effects of such policies – yet they have to do so, or do do so, through the far less effective mechanism of providing property developments.[7] And all this, of course, is in turn overlaid on the more fundamental pattern of an unequal geography of private investment which, while in general terms responding to the logic of the market, also, because of its social form (the deeply conservative nature of financial institutions, the geography of venture-capital houses), persistently fails to respond to commercial opportunities outside the favoured areas. Funding gaps in relation both to property investment and to production have been identified in this chapter. It has been the public sector, in the form of regional bodies and local government, which has shown the way in these areas. Yet their ability to do so, certainly in the case of the latter, is now being restricted by measures from the centre.

NOTES

1 Cited in Invest in Britain Bureau (1986), the same document which markets science parks as attractive locations for multinational capital (see chapter 5).

2 In this chapter we are working with a distinction between 'production', by which we mean the activities of the firms located on the parks, and 'property development'. Property development of course also involves production – the production of parts of the built environment – but the distinction is here between that and other production. This also implies that production in this case means more than direct production or physical production. It includes, for example, research and development.

3 Moreover, there is one member of UKSPA, the Merseyside Innovation Centre, which has property as only a very small element in its overall strategy.

4 Although Henneberry (1986) holds a more optimistic view of science parks from the point of view of developers: 'If science parks can maintain their current development and letting momentum there is certainly the potential for them to satisfy private commercial property development and investment criteria' (p. 5).

5 There is a further problem in that recent legislation seeks to limit local authorities' financial holdings in companies. If the science park is established as a company, therefore, central government legislation potentially limits its financial return and its control and, in turn, the degree of accountability of the science park to the local electorate.

6 These data relate to significant sources of financing. It has been estimated that 25 per cent of Aston firms at one time had some form of government aid (*Venture* 1986: 3). However, this included MSC, other training grants, etc., as well as sources such as the Small Firms Initiative.

7 In the more narrowly defined 'north' of the Assisted Areas, central government regional policy grants are available, and regional agencies – such as the SDA – also may invest in production in various ways. Although such regional assistance has varied greatly over time, the sums can be considerable.

8

CONCLUSIONS

The chapters of this book have approached our subject-matter from a series of different directions. Yet underlying all of them has been a set of common concerns: about the importance of establishing links between different areas of substantive research; about issues of conceptualisation; and, above all, about the policy and wider social and political implications of our arguments.

Under the first heading we have been concerned to establish the importance of the connection between 'science', 'society' and 'space'. At one level this is not new. There are many studies of the geography of new technologies, for instance. Our argument, however, has been different. For too often such studies take the nature of science and technology, and the sociology of their production, as given. Certainly this has been true within geography and in studies of the spatial impact of (scientific and) technological change. One of our aims has been to show, through a detailed exploration of one instance, that this is inadequate. Models of scientific knowledge production and industrial innovation, the social structures of societies and the meaning of those structures, and the spatial organisation of economies and societies, have all varied over time and still do vary between cultures. (What is more – and this in the end is the point we are driving at – there are other, more egalitarian and more socially responsible, alternatives which have as yet barely been tried.) Moreover, and this is fundamental, there is mutual interlinkage between each pair of this threesome, an intimate relation between the forms that each takes and the meanings attributed to them.

Thus the particular, linear, model of scientific production and industrial innovation which has been dominant in the UK, and the particular form which that model has taken, was heavily influenced by the social structure and cultural forms prevalent in the nineteenth century. Further, the model itself has its own social implications, particularly through the nature of the division of labour which is implicit within it. And the symbolism of that was reinforced by the wider social structure and cultural understandings. So social structure was both an influence on and influenced by the dominant model of scientific production and industrial innovation.

The same structure of mutual influences can be drawn out in relation to spatial organisation and spatial form, a particular concern of this book. The science-park case provides an especially clear example. The linear model of science and innovation, at a general level, together with the current form of development of the division of labour within the overall model, has enabled a type of spatial organisation, and specifically spatial separation, not previously possible. In effect it has allowed spatial separation to occur at a different point in the model's sequence of activities. Whereas previously the typical spatial break was between the basic science of academe and the commercial world of production broadly conceived, now the geographical separation can come within the world of production itself, between the shop floor and R&D. This is a phenomenon which grew with the late Fordist spatial separation between different elements of the technical division of labour within major corporations. But science parks have not only further encouraged this move, they have also epitomised it in a wider social sense. The discussion in the Mott report, documented in chapter 6, is an outstandingly clear case of this. For if changes in technology and the division of labour made this new separation possible they did not enforce it. The productive effects of simple spatial proximity between academe and industrial research are not borne out by the evidence, for example. Much more, it seems that the forces which positively produced the spatial separation between R&D and the shop floor were social; and social in the widest sense, which ranges through such issues as the perception of what is a high-status environment, the cachet, still, of having something to do with (certain parts of) academe, and the social clustering which is such an important dynamic of highly specialised labour markets such as these. But, further, the social having produced the change in spatial organisation which the scientific-technological made possible, the new form of spatial organisation then produces its own effects. First, coupled with the given ideological content of the linear model of science and innovation, the spatial separation of R&D from the shop floor contributes to the reformulation of social hierarchies, new bases of social status. The explicitly élitist symbolism of so much of science parks' publicity and physical imagery clarifies and reinforces this effect. And second, the new spatial separation may have its own effects back on the process of industrial innovation. As we have mentioned, the spatial separation of R&D from direct production may be as negative in its real effects on industrial regeneration as spatial proximity of industrial R&D to academe is on its own unfruitful. Viewed in this light, the spatial form of science parks, instead of contributing to increased links between industry and the academic world, may merely be removing one element of the social world of industry to lend it a little of the (fading) glamour of academe. In terms of analysis, the point is that once again the social, the scientific and the spatial are closely linked to each other and mutually influential.

But the question is of course not only one of analysis. It is also intimately related to policy and politics. In a widely cited article, one well known analyst of high-technology industry has written:

> the new industry is likely to be found in regions and in areas quite different from the old. Indeed, the image of the old industrial city – committed to dying industries produced by traditional methods with an ageing workforce resistant to change, with a depressing physical environment that is unattractive to mobile workers, and perhaps lacking the necessary research expertise in the new technologies – is just about as repellent to the new industries as could be imagined. The new industry, then, will seek positively to avoid such places....
>
> We can logically ask whether, in such circumstances, the inner cities in the older industrial regions have any economic future at all. They have an ageing and often outworn infrastructure, which is reaching an age when it demands complete renewal. They have increasingly a residual labour force composed of those who could not or would not join the exodus, and which contains disproportionate numbers of the hard-to-employ. They have a depressing physical environment and often suffer from

> poor transportation linkages compared with suburban and exurban areas. And they suffer from a lack of innovative entrepreneurship ... They lack the right milieu.
>
> They have, in other words, little going for them.
>
> (Hall 1985: 14–15)

Hall goes on to consider what kinds of employment might be found for people in such areas and suggests that the most likely are 'low-paid assembly industry of the Third World type; similarly low-paid work in service industries associated with recreational or tourist developments (conference and exhibition centres, theme parks); some regional-scale office developments, mainly involved in routine data processing (and subject to technological change that may displace some labour and create a demand for special skills which such areas lack). Some such developments, at least, would appear to match demands for certain kinds of labour with the supply of them' (p. 15). The option of creating an R&D tradition, which is interpreted as depending in turn on re-creating an 'entrepreneurial' environment, is seen as fraught with difficulty, most particularly because of the lack of what areas successful in that field almost all possess: 'an exceptionally pleasant environment' (p. 15).

The analysis is both horrifying and, at an immediate level, refreshingly honest. It does not pretend that the regeneration of once dominant but now declining industrial cities and regions can be easily accomplished by the implantation of a few high-tech parks and a revamped regional policy which deals only with the geographical redistribution of already existing jobs. It makes clear, perhaps without fully intending to, the class/social nature of the current spatial distribution of much new industry.

But it is also an analysis which is quite seriously inadequate. Primarily it is so (and apart from more minor criticisms which could be levelled, such as the belief in individual entrepreneurship as the fount of economic growth) because it accepts too much as given. It seems to treat differences in nature between groups of people as innate: the 'hard-to-employ' are contrasted with the environmentally demanding 'mobile workers' as if people were born like that. Similarly, the nature of the new industries, the vast gulf between R&D and 'low-paid assembly industry of the Third World type', is taken as being the product of immutable, presumably technological, necessity. It asks nothing of the nature of 'new industry', of its divisions of labour, nor of the models of scientific production and industrial innovation which lie behind it. It stops short of the really crucial questions which link together science and technology, spatial organisation and the structure of society. It does, however, demonstrate very clearly the consequences if those links are not made. Chapter 6 considered some of the problematical contradictions facing labour-market, and particularly science-park, policies in areas riding the crest of the current form of uneven development and in areas suffering in the wake of previous expansion. In both areas the problems lay in the end in the hierarchical occupational structures of industry and in the social nature of many locational requirements. More widely, throughout the book has run the theme of the North–South divide (or, actually, the rather more complex phenomenon of the contrasts between the outer south and south east of England and much – though not all – of the rest of the United Kingdom). The distribution of scientists and technologists, and the distribution and internal spatial structures of high-tech industry, are only one contributory element in the current form of this long-standing national divide. And yet, we have shown, it is an element unlikely to be solved without a much deeper questioning of the links between spatial form, social structure and, most especially, models of scientific and technological development.

This, then, has been a different kind of policy evaluation, and one in which the attempt has been precisely not to take things as given. For that reason, issues of methodology and of conceptualisation have been crucial. Apart from the points already mentioned, we have sought to go behind the numbers to look at the processes which produce them, and thereby much more seriously to examine the relations of cause and effect postulated (usually implicitly) in the policy itself. In this the definition of concepts has been central. It was the reconceptualisation of science parks themselves which really clarified for us both the ultimate constraints on them as instruments of policy and their wider social implications, the more general issues which they raised. The fact that they are based on a particular model of science and innovation, that they have certain characteristics of spatial form and content, and that they are essentially property developments, is what determines their limits and their possibilities. We are by no means arguing that no science park has had or can have a positive effect in relation to local economic regeneration. A good number of local authorities have turned and twisted and developed and moulded the archetypal model to encourage more progressive policy effects. Apart from the cases, for instance Aston, which we have already discussed, Sheffield's Science Park and the Merseyside Innovation Centre are further clear examples. But the founding assumptions of the archetypal model none the less impose constraints and pose contradictions which are difficult to overcome and which push towards a deeper questioning of the issues at hand.

There has been a continuing thread of concern also with the conceptualisation of difference and of inequality, and what has been striking here is the frequency with which the categories around which science parks are constructed are themselves framed in terms of positive/negative polarities or unequal counterpositions. Many of the characteristics of the structures on which science parks are based are relational in the causal sense used in the earlier chapters of this book – the possession of the characteristic by some actively excludes its possession by others. This has been shown to be the case with the linear model of scientific research and industrial innovation, with the definition of skills in the labour market, and with the social structures to which both these are related. The archetypal science-park model is founded on, and reinforcing of, social inequality.

There has, as we have discussed in the foregoing chapters, for centuries been a mystique about 'science' and those who do it. Certainly our criticisms of the current form of that mystique do not in any sense imply a wish to return to previous forms. As the Conservative governments of the 1980s and early 1990s undermined in the United Kingdom the foundations of the status of the previous technocracy in favour of their own version there has been a tendency to defend 'the professions'. But the professions in some of their guises, and certainly in relation to the world of scientific research, have had their own problems – some of which were illustrated in chapter 6 – of élitism, exclusivity, a widening division of labour, a tendency to think in terms of technocratic 'neutral' solutions, and an assumption that 'we know best'. But none the less, as chapter 5 pointed out, the kinds of changes which have taken place since the late 1970s in the UK have been in reality considerable, have been even further bolstered by a re-fashioning of the mystique on other bases, and, combined with other contemporaneous changes, have grossly exacerbated the social inequality.

There is a fundamental need, for reasons both social and economic, massively to broaden access to science and to technology. This does not mean, as recent UK governments have interpreted it, simply shifting resources within the anyway restricted budget of the educational system. That if anything would produce an even worse result. The issue is much more one of democratising the whole notion of scientific endeavour.

This argument can be made at a number of levels. Most obviously from the point of view of the argument in this book, the inadequacies of the linear model of science and innovation must be challenged. As chapter 3 argued, it is anyway, and perhaps increasingly, inadequate as a descriptive device even in a country such as the United Kingdom where its influence has been strongest. Moreover all the alternative formulations currently gaining in popularity are not only more complex but, in part because of that, less hierarchical and inegalitarian in their implications for the division of labour and its associated social structures. The range of potential alternatives is, however, even wider than this and spans the spectrum from the marginal modification to the really radical rethinking. At the most conservative end of this spectrum the contrasts we have pointed to between, for instance, (West) Germany and the United Kingdom are none the less striking. The assumptions in the former country, certainly in some parts of the economy, about the potential role of shop-floor and production workers more generally in the process of innovation are quite different from in the United Kingdom. The links between industry and academe are also quite differently fashioned. At the other end of the spectrum, if the claims of science to be emancipatory and liberating of human potential are in any full sense to be realised, then the way that science is carried out, and the assumptions about who can participate in its production, must not be founded on inequalities and exclusivities. That means some fairly fundamental re-evaluation. There are still, for instance, some important questions to be answered about scientific procedures and what is considered to be scientific in the West, and the degree to which these are founded on particular, local – especially masculinist – and thereby exclusionary assumptions (Harding 1986; Rose 1983). And, as mentioned in chapter 5, there must remain a future hope that divisions of labour, which must necessarily involve interdependence and therefore a notion of 'the social', need not take a form which is either hierachical or inegalitarian.

There were questions raised by the cultural revolution in China which, though they may not have been answered by that 'revolution', still very clearly need to be posed. Closer to home and more prosaically, the technology networks established by the Greater London Council in the early 1980s were (and some still are) an attempt to broaden the social base and purpose of scientific innovation. Given the minuscule funds at their disposal and their relatively short lives, in comparison with the financing and life-span of the established research institutions, they were only able to hint at an experiment which sorely needs to be taken further (Mole and Elliott 1987). While there clearly needs to be an increase in funding for basic scientific research it does not have to be allocated on an assumption that the potential for innovation and scientific advance lies entirely in the hands of an élite group or on a narrow definition of new 'sunrise' technologies. The technology networks were structured around transport and energy as well as information technology.

But if these visions of the potential futures opened up by a rejection of the linear model seem too radical to be currently contemplated, then at least some of the debilitating peculiarities of the British form of the model could be addressed. Other countries and cultures certainly exhibit their own forms of inequality – the particular élitism of the *techniciens* in France was mentioned in chapter 5 – but the combination of the historical characteristics of the British version of the model with the changes of the last decade or so is producing a particularly acute form of polarisation. The dismal paucity of training in the production workforce generally and the low level of production of scientists and engineers at higher levels are together partly responsible for this. The lack of training of the majority of the work-force effectively disqualifies it from active participation in innovation. At the other end of the skill-training spectrum the shortage, because of underproduction, of more highly qualified workers effectively increases their status and bargaining power. A greatly increased production of such skills would therefore not only be of immense benefit to the economy but would also, by reducing the monopoly power of their holders, be socially egalitarian.

More modestly, but at least equally difficult to achieve in practice, there is the question of the labour process and the design of high-tech jobs themselves (though it is not an issue by any means confined to high-tech jobs). As was documented in chapter 4, this kind of employment all too typically runs directly counter to the grain of movements towards a more equal sharing of paid work, a shorter working day and week and a more equal mixing of paid work with the rest of life. This is not only in itself a dubious change but it imposes constraints on others. As Gorz (1989) has warned, it runs the risk of exacerbating a situation where the economy is structured by a highly paid, high-status group, working all hours at their paid employment, and therefore both needing and being able to command servicing by both unpaid and low-paid labour.

In these ways, then, current developments around 'high technology' – at least as exemplified in science parks – look set to increase the polarisation in society. Yet the effects are negative not only socially but also for the process of industrial innovation and for the economy more widely. And as we have seen, in chapters 6 and especially 7, the existence of an élite group can impose extra costs on production (for instance, through wage costs and through location rents) yet is subsidised through state expenditure.

But as well as social polarisation there are other counter-positions which have also emerged as significant in the analysis. Counterpositions between clean and dirty work environments (closely related to mental and manual work), between the new, involved employee and the old trade-unionised worker 'resistant to change' (see the quotation from Hall earlier), between the (supposed) dynamic activism of the entrepreneurial small company and being closeted in a big, boring corporation or – worse still – the public sector, between the lab-like office in the small town and the factory in the inner city or the north ... between sunrise and sunset. This set of counterpositions, of which the last in some ways encapsulates the flavour of them all, is peculiarly pernicious.

The concept of 'sunrise', especially as used in a dichotomy with 'sunset', is perhaps an unhelpful way of thinking about industries and of dividing up an economy. Sectors which are labelled 'sunrise' cannot be simply detached from other parts of an economic system. Moreover, thinking in these terms tends to encourage a view of technological change which is focused on products rather than also on changes within the production process. In turn by doing that it can lead to a situation where the potential of other sectors, and through that of other areas and regions, is ignored. The textile industry, for example, is rarely thought of as sunrise; it is more usually consigned to the sunset category. Yet recent developments have indicated that the production process even in such an industry can be radically transformed by the application of new forms of technology. This links into the debate, introduced in chapter 6, about the variety of roles which science parks or similar developments can play in a local economy. While in some cases their most appropriate role may be to attempt to introduce new sectors into an economy, in many others the more appropriate strategy may well be the modernisation or transformation of the existing industrial base. It is this latter approach which will usually be more able to draw on the existing strengths of an area, its already established infrastructure, its history of labour skills and developed forms of social organisation. The Japanese examples, of converting older shipbuilding areas into centres of advanced marine engineering, pottery districts into bases for high-technology ceramics and centres of handicrafts into the location for the production of electronic musical instruments, may be instructive here of what is possible with imaginative thinking. Strategies such as these, of course, do require that innovation is closely linked in to the development of changes on the shop floor, and the strict spatial separations of the science-park model become less relevant if not unhelpful.

Japanese experiments in developing broad national technology strategies are important in this context in that longer-term scientific and technology goals have been set for a broad swathe of industries – new and already existing. Included are some with a strong potential direct public dimension, like transport, health and energy as well as the normally cited information technology, biotechnology and new materials.

The fact that the distinction between sunrise and sunset is, or could be, much more blurred than the way in which it is usually conceived is important for another reason. This is quite simply that since in its normal current form the concept of sunrise is restricted to a few sectors, such as electronics, information technology and biotechnology, there is not enough to go round. It would simply be impossible, for instance, for all the science parks in the United Kingdom to succeed in making their local economies serious centres of such sunrise industries. Here again conceptualisation is important; not just in the definition of sunrise and sunset, but in the idea that because a particular set of characteristics 'worked' in, say, Stanford and Silicon Valley their reproduction elsewhere will reproduce also their results. This is a fallacy. In part it is so because, even were the local-level conditions to be reproduced (which in itself is unlikely), the wider situation has changed. Some areas are now already established and the dynamics of location have consequently shifted. While the actual geographical location of clusters of new industry may be strongly based on social reasons, as indicated in the argument above (they do not all have to be in 'environmentally pleasing' areas, for instance, or in the south east of England within the UK), there do seem to be some real economic forces pushing towards a degree of clustering itself. These run from the clear feeling of a need to be 'in on the scene' to the problems of acquiring particular forms of skilled labour. The pull of some of these factors could be reduced, for instance through increased training, as mentioned above. But the tendency to some degree of clustering is unlikely to be eradicated. This, however, strengthens still further the argument against a proliferation of small, local attempts to generate a new 'high-technology' sector. In the UK the better way to dilute the concentration in Cambridge and in the south and east more generally might be to establish a few major centres which could really function as counter-magnets. There are large well established academic and industrial centres in the north which could serve as focuses, especially if the definition of sunrise were to be expanded in the ways already suggested. Moreover, the example of Cambridge indicates that this would require considerable and sustained public support from central government funding. The trick would be to do this without generating social conflicts. If the new breed of scientists and technologists and venture capitalists, whose conception of their own high status has been indicated by many a quotation in this book, are likely to look coldly at the prospect of moving to, say, Yorkshire or Lancashire (though in fact the latter was and is one of their most crucial bases in an older social form), then neither is it the case that the resuscitation of 'the north' can be accomplished by the implantation of islands of middle-class culture. The increased funding of the existing major bases of science and technology outside the south and east should go along with an attempt to shift the social structure of the whole process of scientific production and technological innovation – in effect to democratise it, to move away from the linear model.

And yet, as we have seen throughout this book, much of what lies behind the glaring social polarisation and the hierachical counterpositions is fantasy or bolstered by fantasy. The heroic individual scientist-inventor is not really the crucial element in scientific advance, nor should be. The level of science and technology on science parks falls considerably below the claims of much of the rhetoric. The numbers of scientists-turned-entrepreneur are relatively few, the predominance of small companies is very greatly exaggerated. The endless excited quotations from journalists, the overall hype of high tech, are frequently overdone. Yet fantasies, too, have their effects. In this case, and especially because they are so much based on status, on comparison, on counterposition against others, one important effect of the fantasy is further to exacerbate the inequality and the polarisation already implicit in more material processes. One thing which is definitely a fantasy is that the hype (and the real benefits of participation) of high tech as it currently exists can be generalised to everyone. By its very constitution, material and rhetorical, it cannot, and for that reason alone it should be challenged. The fact that it is also unproductive for the economy and in the end for any socially progressive science and technology in society as a whole only reinforces the case.

REFERENCES

Advisory Conciliation and Arbitration Service (1988) *Labour flexibility in Britain: the 1987 Acas survey*, Acas Occasional Paper No. 41, London: ACAS.

Agricultural Genetics Company Ltd (1986) *A Unique Opportunity*, Cambridge: AGC.

Armstrong, Peter (1987a) *The Abandonment of Productive Intervention in Management Teaching Syllabi: an historical analysis*, Warwick Papers in Industrial Relations 15, Industrial Relations Research Unit, University of Warwick.

Armstrong, Peter (1987b) 'Engineers, management and trust', *Work, Employment and Society*, 1, 4: 421–40.

Association of Graduate Careers Advisory Services (1988) *University Graduates: summary of first destination and employment*, AGCAS.

Association of Scientific Workers (1947) *Science and the Nation*, Harmondsworth: Penguin.

Association of University Teachers (1990) *Goodwill under Stress*, London: AUT.

Aston Science Park (1986) *Venture*, 1, 1, November.

Atkinson, J. (1984) 'Flexibility: Planning for an uncertain future', *IMS Manpower Review*, 1, 1, August.

Bain, G. S. (1983) *Industrial Relations in Britain*, Oxford: Blackwell.

Becker, U. (1989) 'Class theory and the social sciences: Erik Olin Wright on classes', *Politics and Society*, 17, 1: 67–88.

Beckett, Sir Terence (1987) 'Innovation in the twenty-first century', Institute of Electrical Engineers (IEE), *Electronics and Power*, London, 33, 1: 26–30.

Bernal, J. D. (1969) *Science in History*, Vol.2, Harmondsworth: Penguin.

Bijker, W. E., Hughes, T. P., and Pinch, T. J. (1989) *The Social Construction of Technological Systems*, Cambridge, Mass., and London: MIT Press.

Birmingham City Council (1982) *Economic Bulletin*, 1, June.

Birmingham City Council (1985) 'Science Park Working Party Report', Economic Development Unit, 26 June.

Birmingham City Council (1986) *1986 Review*.

Birmingham City Council (1990) 'Aston Science Park: Economic Impact Study', Economic Development Committee.

Blank, S. (1973) *Industry and Government in Britain: the Federation of British Industry in politics, 1945–65*, Farnborough, Hants., and Lexington, Mass.: Saxon House/Lexington Books.

Boddy, M., Lovering, J., and Bassett, K. (1986) *Sunbelt City? A study of economic change in Britain's M4 growth corridor*, Oxford: Clarendon Press.

Bolton, W. K. (1986) *Entrepreneurial Opportunities for the Academic*, Paper presented to the UKSPA Annual Conference, London, 11 December.

Bond, C. P. (1985) 'Targetting a science park to its task and market', in Gibb, J. M. (ed.) *Science Parks and Innovation Centres: their economic and social impact*, Amsterdam: Elsevier.

Bradfield, J. (1983) 'Introduction' to *Cambridge Science Park Directory*, Cambridge: CSP, January.

Braverman, Harry (1974) *Labour and Monopoly Capital: the degradation of work in the twentieth century*, New York and London: Monthly Review Press.

Briggs, A. (1968) *Victorian Cities*, Harmondsworth: Penguin.

Brindle, D. (1987) 'R&D technologists "hit salary bar", survey finds', *Financial Times*, 3 July 1987.

British Telecom (1986, 1987, 1989) *Report and Accounts*.

British Telecom (1989) *Supplementary Report*.

British Telecom (1990) *Annual Review*.

Burns, T., and Stalker, G. M. (1961) *The Management of Innovation*, London: Tavistock.

Buxton, A. (1988) 'Where dishevelled dons are giving way to boffins with more BMWs than bikes', *The Guardian*, 5 September.

Buxton, James (1987) 'Scottish electronics – spin-offs for entrepreneurs', *Financial Times*, 31 March: 38.

Cambridge City Council (1986) *Employment Development Strategy: high technology and conventional manufacturing industry*, Environment Committee, November.

Cambridge Local Collaborative Project (1986) *Training needs of new technology industries in the Cambridge area: the final report*, Cambridge: CLCP.

Cambridge University Reporter (1969) 'The relationship between the university and science-based industry' (The Mott report), 22 October.

Carchedi, G. (1977) *The Economic Identification of Social Classes*, London: Routledge & Kegan Paul.

Carter, N., and Watts, C. (1984) *The Cambridge Science Park*, London: Surveyors Publications.

Carter, R., and Kirkup, G. (1990) *Women in Engineering: a good place to be?* Basingstoke: Macmillan.

Chandler, A. D., Jr. (1977) *The Visible Hand*, Cambridge, Mass.: Harvard University Press.

Cockburn, C. (1985) *Machinery of Dominance: women, men and technical know-how*, London: Pluto Press.

Connor, H., and Pearson, R. (1986) *Information Technology Manpower into the 1990s,* Brighton: Institute of Manpower Studies.

Cookson, C. (1982) 'Not exactly a park, rather a high-tech nursery', *The Times,* 1 July: 11.

Cooley, M. (1972) *Computer Aided Design – its nature and implications,* Richmond, Surrey: Associated Union of Engineering Workers, Technical and Supervisory Section.

Cooley, M. (1980) *Architect or Bee? The human/technology relationship,* Slough: Langley Technical Services.

Cooper, A. C. (1971) 'Spin-offs and technical entrepreneurship' *IEEE Transactions on Engineering Management, EM–18,* 1: 2–6.

Cornelius, A. (1989) 'Graduates in engineering shun profession', *The Guardian,* 8 December.

Dalton, I. (1985) Opening Statement by the Chairman of the UKSPA, Annual Conference, December.

Danson, M., Fairley, J., and Kerevan, G. (1987) 'Deconstructing the Scottish economy: a critical assessment of the Scottish Development Agency and its technology strategy', paper presented at the South East Economic Development Strategy Conference on Technology, Brighton, 12 March.

David, R. (1986) 'Dilemma of balancing development', *Financial Times,* 21 July.

Davies, Brian (1988) 'Physics in higher education – a degree of uncertainty?' *Physics Bulletin,* 39, 2: 63–4.

Davies, P. (1989) 'Let them eat crumbs', *The Guardian* (Education Supplement), 9 May.

Debenham Tewson & Chinnocks (1983) *High Technology: myths and realities: a review of developments for knowledge-based industries,* London: Debenham Tewson & Chinnocks, chartered surveyors, July.

Debenham Tewson & Chinnocks (1987) *Industrial Rent and Rates 1987,* London: Debenham Tewson & Chinnocks.

Dixon, M. (1987) 'How top specialists are ranked across Europe', *Financial Times,* 27 May: 18.

Dixon, M. (1988) 'How jobs market reacted to Black Monday', *Financial Times,* 27 January.

Dixon, M. (1990) 'Bleakest summer for hunters since 1981', *Financial Times,* 31 October.

Dodsworth, T. (1988) 'New attitudes to break the IT deadlock', *Financial Times,* 19 December.

Dorfman, N. S. (1983) 'Route 128: the development of a regional high technology economy', *Research Policy,* 12: 299–316.

Dosi, G. (1988) 'The nature of the innovation process', in Dosi *et al.* (1988): 221–38.

Dosi, G., Freeman, C., Nelson, R., Silverberg, G., and Soete, L. (1988) *Technical Change and Economic Theory,* London: Frances Pinter.

Elson, D. (1987) 'Market socialism or socialization of the market?' Draft one, mimeo.

Engineering Council (1985) *The 1985 Survey of Professional Engineers,* London: Engineering Council.

Engineering Council (1987) *The 1987 Survey of Professional Engineers*, London: Engineering Council.

Engineering Industry Training Board (1986) *Occupational Profile: trends in employment and training of professional engineers, scientists and technologists in the engineering industry*, Stockport: EITB Publications.

Engineering Industry Training Board (1989) *Economic Monitor*, No. 30.

Eustace, P. (1985) 'Cambridge: the in-place in high technology', *The Engineer*, 14 February: 20–1.

Eustace, P. (1989) 'Who wants to be an engineer? You don't. *The Engineer* survey', *The Engineer*, 6 April: 23.

Fazey, I. H. (1987) 'Stark contrasts between regions', *Financial Times*, 19 January.

Financial Times (1981) 'Science and industry find a matchmaker', 24 November.

Financial Times (1984) 'Cambridge – the house of high tech', 23 August.

Financial Times (1984) 'High-tech companies that worry about neighbours', 12 October.

Financial Times (1985) 'All manufacturing excluded', in *Science Parks Weekend Report*, 30 November: xii.

Financial Times (1987), 31 March.

Financial Times (1988) 'Research workers see sharp rise in pay', 23 August.

Foster, K. (1982) 'The Job Generation Process', mimeo of lecture given 14 June, Birmingham.

Freeman, C. (1982) *The Economics of Industrial Innovation*, 2nd edition, London: Frances Pinter.

Fuller Peiser (1985) *High Technology' 85*, London: Fuller Peiser.

Gaffikin, F., and Nickson, A. (1984) *Jobs Crisis and the Multinationals: de-industrialisation in the West Midlands*, Birmingham: Birmingham Trade Union Resource Centre.

Garnett, N. (1989) 'No way to treat young engineers, says survey', *Financial Times*, 8 December.

Glover, I. A., and Kelly, P. (1987) *Engineers in Britain: a sociological study of the engineering dimension*, London: Allen & Unwin.

Gorz, A. (1967) *Strategy for Labour: a radical proposal*, Boston: Beacon Books.

Gorz, A. (1989) *Critique of Economic Reason*, London: Verso.

Guardian (1985) 'The slimline tonic', 5 March.

Hacker, Sally (1981) 'The culture of engineering: woman, workplace and machine', *Women's Study International Quarterly*, 4: 341–53.

Hall, P. (ed.) (1981) *The Inner City in Context: the Final Report of the Social Science Research Council Inner Cities Working Party*, London: Heinemann Educational.

Hall, P. (1985) 'The geography of the fifth Kondratieff', in Hall, P., and Markusen, A. (eds.) *Silicon Landscapes*, London: Allen & Unwin: 1–19.

Hall, P., Breheny, M., McQuaid, R., and Hart, D. (1987) *Western Sunrise*, London: Allen & Unwin.

Harding, S. (1986) *The Science Question in Feminism*, Ithaca, N.Y.: Cornell University Press.

Henneberry, J. M. (1984a) *A Survey of British Science Parks and High Technology Developments*, Sheffield: PAVIC.

Henneberry, J. M. (1984b) 'Property for high-technology industry', *Land Development Studies*, 1: 145–68.

Henneberry, J. M. (1986) 'Science parks: an evaluation', Centre for Local Economic Strategies seminar, 16 April.

Hounsell, D. A., and Smith, J. K., Jr. (1988) *Science and Corporate Strategy: Du Pont R&D 1902–1980*, Cambridge and New York: Cambridge University Press.

Ince, Darrell (1988) 'Five years on', *Datalink*, 15 February: 12.

Ince, M. (1989) 'Shifting sands of scientific research', *The Times Higher Education Supplement*, 3 November.

Incomes Data Services (1986) *Flexibility at Work*, IDS Study 360, London: Incomes Data Services.

Incomes Data Services (1987) *Pay and Progression in Research and Development*, London: IDS Top Pay Unit.

Incomes Data Services (1988) *Skill Shortages in the South East: engineering*, IDS Report Labour Market Supplement, No. 2, March: 4–8.

Information Technology Skills Agency (1988) *Changes in IT Skills*, available from Stewart Judd, CBI, Centre Point, 103 New Oxford Street, London WC1A 1DU.

Institute for Employment Research (1989) 'Occupational studies', *Bulletin*, 3:1, Institute for Employment Research, University of Warwick.

Invest in Britain Bureau (1986) *UK Science Parks*, London: IBB.

IVL (1989) *Survey into the Attitudes and Intentions of Graduating Engineers*, London: IVL.

Kaplinsky, R. (1989) 'Restructuring the capitalist labour process: implications for administrative reform', *IDS Bulletin*, 19, 4: 42–9.

Kelly, P., Kransberg, M., Rossini, F., Baker, N., Tarpley, F., and Mitzner, M. (1978) 'Introducing innovation', in Roy and Wield (1986).

Kehoe, L. (1987) 'A U-turn in Silicon Valley', *Financial Times*, 9 February.

King, A. (1974) *Science and Policy: the international stimulus*, London: Oxford University Press.

Kline, S. (1985) 'Research is not a linear process', *Research Management*, 28, July–August.

Kline, S. J. (1989) *Innovation styles in Japan and the United States: cultural bases; implications for competitiveness*, Report INN-3B, Thermosciences Division, Mechanical Engineering, Stanford University, December.

Labour Research Department (1986) *Bargaining Report* No.56, November.

Larsen, J. K., and Rogers, E. M. (1984) *Silicon Valley Fever*, Hemel Hempstead: Unwin Hyman; New York: Basic Books.

Leadbeater, C. (1988) 'Britain badly outpaced in training courses', *Financial Times*, 21 November.

Leadbeater, C. (1989) 'Long-term staff shortages likely in information technology sector', *Financial Times*, 19 January.

Levi, P. (1985) 'Town, gown and 10 miles around: Cambridge keys in', *The Times*, 12 February.

Little, A. D. (1913) 'Industrial research in America' *Journal of Industrial and Engineering Chemistry*, 5: 793.

Lloyd, J. (1986) 'A trinity of trading, growing and making', *Financial Times*, 17 March.

Macdonald, S. (1983) 'Technology beyond machines', in Macdonald, Stuart, Lamberton, D. McL., and Mandeville, Thomas, *The Trouble with Technology*, London: Frances Pinter.

MacKenzie, D., and Wajcman, J. (1985) *The Social Shaping of Technology*, Milton Keynes: Open University Press.

MacPherson, D. (1985) 'The new glittering prizes', *Sunday Times*, 27 January.

Mallet, Serge (1975) *The New Working Class*, Nottingham: Spokesman.

Marquand, J. (1979) *The Service Sector and Regional Policy in the United Kingdom*, Centre for Environmental Studies Research Series, No. 29.

Marsh, P. (1984) 'A hustler in the techno park', *Financial Times*, 10 October: 23.

Marsh, P. (1985) 'Hi-tech entrepreneurs top the bill in Cambridge success story', *Financial Times*, 11 February.

Marsh, P. (1986a) 'High-tech threat to Cambridge tranquillity', *Financial Times*, 27 May.

Marsh, P. (1986b) 'A dream they never sold', *Financial Times*, 9 June.

Marsh, P. (1986c) 'Elusive ingredients of a high growth recipe', *Financial Times*, 13 June.

Marsh, P. (1987) 'Trauma is sometimes good for you', *Financial Times*, 11 February.

Marshall, M. (1985) 'Technological change and local economic strategy in the West Midlands', *Regional Studies*, 19, 6: 570–8.

Marshall, M. (1987) *Long Waves of Regional Development*, Basingstoke: Macmillan.

Mason, C. (1987) 'Venture capital in the United Kingdom: a geographical perspective', *National Westminster Bank Quarterly Review*, May: 47–59.

Massey, D. (1984) *Spatial Divisions of Labour*, London: Macmillan.

Massey, D. (1988) 'Uneven development: social change and spatial divisions of labour', in Massey, D., and Allen, J., *Uneven Re-Development*, London: Hodder & Stoughton, in association with the Open University.

Massey, D., and Meegan, R. (1979) 'Labour productivity and regional employment change', *Area*, 11, 2: 137–45.

Meacher, Michael (1987) 'Why Labour must extend its class appeal', *The Guardian*, June 13.

Mole, V., and Elliott, D. (1987) *Enterprising Innovation: an alternative approach*, London: Frances Pinter

Monck, C. S. P., Quintas, P., Porter, R.B., Storey, D. J., and Wynarczyk, P. (1988) *Science Parks and the Growth of High Technology Firms*, London: Croom Helm.

Moore, B., and Spires, R. (1983) *The Experience of the Cambridge Science Park*, Research Technology and Regional Policy Workshop, 24–7 October, Paris, OECD.

Moreton, A. (1986) 'First to do the double', *Financial Times*, 21 July.

Morgan, B. (1986) 'Parking space', *Times Higher Education Supplement*.

Morgan, K., and Sayer, A. (1988) *Microcircuits of Capital*, Oxford: Polity.

Mowery, D. (1983) 'The relationship between intra-firm and contractual forms of industrial research in American manufacturing, 1900–1940', *Explanations in Economic History*, 20, 4: 351–74.

Mowery, D. C. (1986) 'Industrial research, 1900–1950', in Elbaum, B., and Lazonick, W. (eds.) *The Decline of the British Economy*, Oxford: Clarendon Press: 189–222.

National Economic Development Office (1986) *Changing Working Patterns*, NEDO.

National Economic Development Office (1989) *Switching on Skills*, NEDO.

Nelson, R. R. (1989) 'What is private and what is public about technology?' *Science, Technology and Human Values*, 14, 3: 229–41.

New Scientist (1987), 'Royal Society plumbs the brain drain', 2 July: 23.

Noble, D. (1984) *Forces of Production: a social history of industrial automation*, New York: Knopf.

Noble, David (1989) 'The castration of Abelard and the culture of science', presentation to Society for the History of Technology Annual Conference, Sacramento, October.

Oakey, R. (1985) 'British university science parks and high-technology small firms: a comment on the potential for sustained industrial growth', *International Small Business Journal*, 4, 1: 58–67.

Page, I. (1985) 'The role of the university and its science park to Bradford's economic strategy', paper to UKSPA Conference, London, December: 1.

Pavitt, K. (1982) 'R&D, patenting and innovative activities: a statistical exploration', *Research Policy*, 11, 1: 33–51.

Pavitt, K. (1984) 'Sectoral patterns of technical change: towards a taxonomy and a theory', *Research Policy*, 13: 343–73.

Pender, T. (1985) 'Science parks – getting the property component right', paper presented to UKSPA Conference, December.

Peters, L., and Fusfeld, H. (1982) 'Current US university–industry research connections', in National Science Foundation, *University–Industry Research Relationships*, Washington, DC: NSF.

Pettigrew, A. M. (1985) *The Awakening Giant: continuity and change in Imperial Chemical Industries*, Oxford: Basil Blackwell.

Pinch, T., and Bijker, W. E. (1989) 'The social construction of facts and artifacts: or how the sociology of science and the sociology of technology might benefit each other', in Bijker, W. E., Hughes, T. P., and Pinch, T. J., *The Social Construction of Technological Systems*, Cambridge, Mass., and London: MIT Press.

Piore, M. J., and Sabel, C. T. (1984) *The Second Industrial Divide*, New York: Basic Books.

Pollert, A. (1988) 'Dismantling flexibility', *Capital and Class*, 34: 42–75.

Poulantzas, N. (1975) *Classes in Contemporary Capitalism*, London: New Left Books.

Prais, S. J. (1981) 'Vocational qualifications of the labour force in Britain and Germany', *Economic Review*, 98.

Prais, S. J., and Wagner, K. (1983) 'Some practical aspects of human capital investment: training standards in five occupations in Britain and Germany', *Economic Review*, 105.

Pryke, M. (1989) 'Urban land values and the changing role of financial institutions: a case study of the City of London,' unpublished PhD thesis, Open University.

Rees-Mogg, William (1987) 'A revolution that is sweeping away the cause of socialism', *The Independent*, 5 May.

Reward Research Group (1988) *The Research and Development Survey, 1988*, Reward, 1 Mill Street, Stowe, Staffordshire ST1F 8BA.

Richmond, M. H. (1985) 'Our objectives for Manchester Science Park', paper to UKSPA Annual Conference, December.

Rimmer, A. (1985) 'Management and financial support for innovation: the experience of the Merseyside Innovation Centre', paper presented to UKSPA Annual Conference, December.

Roberts, E. B., and Wainer, H. A. (1968) 'New enterprises on Route 128', *Science Journal*, 4, 12: 78–83.

Rogers, R. (1985) 'The slimline tonic', *The Guardian*, 5 March.

Rose, H. (1983) 'Hand, brain and heart: a feminist epistemology for the natural sciences', *Signs: Journal of Women in Culture and Society* 9, 1.

Rose, H., and Rose, S. (1970) *Science and Society*, Harmondsworth: Penguin.

Rose, H., and Rose, S. (1982) 'The 1964–70 Labour government and science and technology policy' (notes for the Group for Alternative Science and Technology Strategy).

Rowe, D. (1985) 'Financing the property component – the experience of the University of Warwick Science Park in obtaining public and private sector finance', UKSPA Annual Conference, London.

Roy, R., and Bruce, M. (1984) 'Product design, innovation and competition in British manufacturing: background aims and methods', Working Paper WP-O2, Design Innovation Group, Open University, September.

Roy, R., and Wield, D. (1986) *Product Design and Technological Innovation*, Milton Keynes and Philadelphia: Open University Press.

Royal Society (1987) *The Migration of Scientists and Engineers to and from the UK.*

Rudge, A. (1986) 'Research, development and decline: Britain's industrial enigma?' *Electronics and Power*, May, 347–52.

Savage, M., Dickens, P., and Fielding, T. (1988) 'Some social and political implications of the contemporary fragmentation of the "service class" in Britain', *International Journal of Urban and Regional Research*, 12, 3: 455–76.

Saxenian, A. L. (1983) 'The urban contradictions of Silicon Valley: regional growth and the restructuring of the semiconductor industry', *International Journal of Urban and Regional Research*, 7, 2: 237–61.

Saxenian, A. L. (1985) 'Let them eat chips', *Environment and Planning* D: *Society and Space*, 3: 121–7.

Saxenian, A. L. (1989) 'In search of power: the organisation of business interests in Silicon Valley and Route 128', *Economy and Society*, 18, 1: 25–70.

Sayer, A. (1984) *Method in Social Science: a realist approach*, London: Hutchinson.

Scottish Development Agency (1984) *Annual Report 84*, Glasgow: SDA.

Segal, N. (1985) 'The Cambridge Phenomenon', *Regional Studies*, 19, 6: 563–70.

Segal Quince & Partners (1985) *The Cambridge Phenomenon*, Cambridge: Segal Quince & Partners.

Senker, P. (1991) 'Skill Shortages and Britain's competitiveness', in Bosworth, D., *et al.* (1991) *Skill Shortages*, Macmillan, forthcoming.

Shaw, John (1984) 'A high-tech check for the "Cambridge Phenomenon"', *Daily Telegraph*, 27 February.

Shaylor, G. (1985) *An Economic Strategy for Birmingham 1985/86*, Birmingham City Council.

Silberston, A. (1987) *Is there a shortage of engineers?*, Papers in Science, Technology and Public Policy, 16., London: Imperial College.

Simpson, M., and Smith, P. (1986) 'Where the jobs are', *New Scientist*, 7 August.

Smith, C. (1987) *Technical Workers: class, labour and trade unionism*, Basingstoke: Macmillan.

Smith, T. (1987) 'The three year cycle that ends when you're thirty', *The Guardian*, 16 April.

Steedman H. (1988) *Vocational Training in France and Britain: mechanical and electrical craftsmen*, National Institute Discussion Paper No. 130, London: NIESR.

Steward, F., and Wield, D. (1984a) 'Science Planning and the State', D209, *The State and Society*, Milton Keynes: Open University.

Steward, F., and Wield, D. (1984b) 'Science planning and the state' in McLennan, Gregor, Held, David, and Hall, Stuart (eds.) *State and Society in Contemporary Britain*, Cambridge: Polity Press.

Storey, D. J., and Strange, A. (1990) 'Where are they now? Some changes in firms located on UK science parks in 1986', paper presented to the Fifth Annual Conference of UKSPA, Birmingham.

Swords-Isherwood, Nuala (1980) 'British management compared', in Pavitt, Keith (ed.) *Technical Innovation and British Economic Performance*, London: Macmillan.

Tarsch, J. (1985) *Graduate shortages in science and engineering*, Department of Employment Research Paper No. 50, London: DoE.

Taylor, A. (1987) 'Graduates "lured" away from engineering jobs', *Financial Times*, 12 February.

Taylor, C., and Silberston, A. (1973) *The Economic Impact of the Patent System*, Cambridge: Cambridge University Press.

Teece, D. J. (1986) 'Profiting from technological innovation: implications for integration, collaboration, licensing and public policy', *Research Policy*, 15: 285–305.

Teece, D. (1988) 'Technological change and the nature of the firm', in Dosi *et al.*

Thompson, E. P. (ed.) (1970) *Warwick University Ltd*, Harmondsworth: Penguin.

Thrift, N. (1987) 'Sexy Greedy: the new international financial system: the

City of London and the South East of England', University of Bristol Papers on Producer Services, No. 7, Department of Geography, University of Bristol.

Times Higher Education Supplement (1988), '"Anything goes" to correct computer skills shortage', 18 March.

The Times (1985) 'London's first science park', 9 December.

UKSPA (1985) 'Foreword' to *Science Park Directory*, Sutton Coldfield: UKSPA.

UKSPA (1987) *Tenants Directory*, Sutton Coldfield: UKSPA.

UKSPA (1989) 'Summary of Operational Parks in the UK', paper presented to Fifth Annual UKSPA Conference, Edinburgh, April.

University Funding Committee (1990) *University Statistics 1988–89*, Cheltenham: University Statistical Record.

University Grants Committee (1982) *University Statistics 1980–81*, Cheltenham: University Statistical Record.

Vig, N. T. (1986) *Science and Technology in British Politics*, Oxford: Pergamon Press.

Vincenti, W. G. (1984) 'Technological knowledge without science: the innovation of flush riveting in American Airplanes, ca 1930–ca 1950', *Technology and Culture*, 25: 540–76.

Virgo, Philip (1987) *The IT Skills Crisis*, Manchester: National Computing Centre.

von Hippel, E. (1988) *The Sources of Innovation*, New York: Oxford University Press.

Ward, M. (1987) *The Basildon Economy: growth and prospects*, Basildon Economic Development Corporation Ltd.

Watson, H. B. (1976) 'Organisational bases of professional status: a comparative study of the engineering profession', unpublished PhD thesis, University of London.

West Midlands County Council Economic Development Committee (1984) *Economic Review No. 1.*, Birmingham: WMCC.

West Midland Regional Strategy Review (1985) 'Employment and Economic Regeneration', Background Paper No. 4, February.

Whalley, P. (1986) *The Social Production of Technical Work: the case of British engineers*, Basingstoke: Macmillan.

Whalley, P., and Crawford, S. (1984) 'Locating technical workers in the class structure', *Politics and Society*, 13, 3: 239–52.

Wiener, M. (1985) *English Culture and the Decline of the Industrial Spirit, 1850–1980*, Harmondsworth: Penguin.

Williams, N. (1990) 'Britain lags £10bn in research funding: figures put Germany far ahead on pay and equipment', *The Guardian*, 2 October.

Wright E. O. (1976) 'Class boundaries in advanced capitalist societies', *New Left Review*, 98, July–August: 3–41.

Wright, E .O. (1978) *Class, Crisis, State*, London: New Left Books.

Wright, E. O. (1985) *Classes*, London: Verso.

Wright E. O. (1988) 'Exploitation, identity and class structure: a reply to critics of *Classes*', mimeo.

Wyatt, S. (1985) 'The role of small firms in innovative activity: some new evidence' *Economia e Politica Industriale*, 45: 47–82.

INDEX

absenteeism 111–12
academe–industry links 7, 60–72 *passim*, 162, 195–6; as aim of parks 22–3, 28, 34–40, 72–6, 168; away from park 74; criterion for 14; and national research laboratories 196; social aspects of 36, 40, 121, 196; spatial aspects of 74–5, 159, 160, 241
academic start-ups 35–8, 72
access to science and technology 245–6
accumulation, rounds of 197–8, 203–4
activities of firms 45–6, 186; and public-sector investment 226–35
Advanced Manufacturing Centre 186
age of firms 33
agglomeration 156–60, 184–7
Agricultural Genetics Company Limited 233, 235
Agriculture and Food Research Council 45, 233
aims and objectives of parks 8, 21–9, 224–6, 237; evaluation of 30–50
alternative conceptualisation 9, 158–62, 213, 243–4; and linear model 60, 158–9
apprenticeships 122–3, 140, 145, 150
appropriation of labour 126–7, 128–9
Armstrong, Peter 69
assets, ownership of 126–30; *see also* skills and credentials
Aston Science Park:
academe–industry links at 36–7, 38, 74, 195–6; aims of 25–7; employment at 94, 101, 110, 187–90, 193–4; (job creation 41; shortages 154, 193–4); international links of 178–84 *passim*; level of technology at 45, 181; and local economy 165–96 *passim;* new start-ups at 32; property of 168–9, 170, 177, 222; reasons for locating at 161; setting of 163–4; sources of finance for 228, 229, 232–5
Aston University 169–70
autonomy: concept of 124; in work practices 91–107; *see also* flexibility

Barclays Bank 19, 220
BASF 61
Bayer 61
Beckett, Sir Terence 75
Belgium 137
Bernal, J.D. 61–2, 173
Bioscot Ltd 203
Biotechnology 44, 181, 200, 203
Birchwood Science Park 14
Birmingham: economy of 164, 165–7, 187–8; *see also* Aston Science Park
Birmingham City Council 25–6, 168–9, 186–7, 196

Birmingham Research Park 196
Birmingham Technology Ltd 26–7, 36, 170, 229
Blackett, Patrick 69, 173
Blank, S. 67–8
BOC 19, 23, 214
Bolton, Dr Bill 39, 144
bonus schemes 103
Bradfield, Dr John 24
Bradford City Council 21–2
Bradford Science Park: academe–industry links at 37; aims of 21–2; new start-ups at 32; rents at 223
branch-plant 200
'brain drain' 132–3
Braverman, Harry 116
British Dyestuffs 65
British Nuclear Fuels Ltd 14
Brunel Science Park 17
Burns, T. 62–3, 120
Business in Liverpool Ltd 32
Buxton, A. 144

Cambridge 89; economy of 164, 175, 191–2
Cambridge Science Park 17; academe–industry links at 35–40 *passim,* 195–6; aims of 24–5; exclusivity at 110, 111; freelance employees at 101; incentives at 103; international links of 178–84 *passim;* job creation by 41; level of technology at 44–5, 181; and local economy 170–96 *passim;* and local labour market 190–4; new start-ups at 32; property of 177, 214, 215; public presentation of 87–91, 144, 164, 182; shortages at 154, 193–4; sources of finance for 228, 229, 232–5, 237; trade unions at 105; training at 146; working hours at 94, 95
Carchedi, G. 116
career structure 117–18, 122–3, 131, 132–3, 138
Carter, N. 24–5
CBI (Confederation of British Industry) 75
Chandler, A.D., Jr 64
China 85, 246
class and status 3–4, 11; and industry 7, 59, 62, 122–3; in public sector 150–1; as relational concept 124–30 *passim,* 146; of scientists and technologists 115–62; and social power 148–62
clustering of science parks 249
competition 208
concentration 210
conceptualisation of science parks 9–10, 243; alternative 9, 60, 158–62, 213, 243–4; popular 9, 13–51, 71–2; *see also* image; relational characteristics
Connor, H. 152–3
Contact 32
Cooley, M. 116
counterposition of science parks with rest of economy 87, 107, 113–14, 150–1, 160
Coventry City Council 19, 221
crafts 140–1
Crawford, Sir Frederick 26, 169
Crawford, S. 122–3
credentials 126–30, 140–1, 145–6
customers: location of 179, 181, 185, 187; public-sector 232; in Scotland 200–1, 202

data-processing 140, 142
David, R. 92
Davies, B. 135, 152
Davies, P. 157
decision-making: participation in 70, 97–9; over pay 103–4
definition of a science park 13–14
Delta Metals 170, 184, 186
design of science parks 87–91, 160–1
Development Councils 67–8
diffusion *see* innovation
division(s) of labour 3–4, 7, 10, 11, 245–6; and flexibility 96, 111–13; and innovation 80; in linear model 58–9, 60, 85; and specialisation 71; and time

monitoring 96; *see also* employment
Dixon, M. 137
Dosi, G. 79–83
Dunlop 186
dyestuffs industry 60–2, 65

Edinburgh 200, 202
Edwards, Christopher 217
electronics 43–5, 140–3, 201
Elson, D. 114
emigration of scientists and technologists 132–3
employment 5, 11, 19–20, 76, 86–114, 241–3; in academe 130–3; casual 106; clerical 98, 110; consultancy 96; creation of 28, 40–3, 168, 188–9; female 43, 110–11, 112, 149–50, 188, 191; flexibility in 92–114; freelance 101–2, 106; future of 87; high-status 130–7, 156–8; imagery of 86–91, 107, 113–14, 160–1, 206–9, 241, 250; and local economy 187–94; manual 110, 192, 194; and North–South divide 17, 19–20, 153–6, 175–6; part-time 112, 192; semi-skilled 111, 192, 194; shortages 146–7, 151–6, 188, 193–4, 246; social aspects of 76, 115–62, 173–4, 189–93, 204, 246–7; spatial aspects of 154–8, 189–94, 207–9; unskilled 110, 194; *see also* division of labour; qualifications
engineers 138, 140–3
English Estates 22, 217, 219
Enterprise Zones 236
entrepreneurship 141–3, 148, 250; *versus* public sector 206–9
Environmental Protection Unit laboratories (Aston) 186–7
ethnic minorities 149, 188
Eustace, P. 138
exclusivity 109–14, 159–61, 223, 244
expansion of firms 41
exploitation 126–7, 128–9

failure of firms 32–3
Ferranti 65, 200
Ffowcs-Williams, Professor John 90
financial sector 134, 152; investment in science parks by 221–2, 225
flagship companies 183–4
flexibility 92, 106–7; and exclusivity 109–14; organisational and functional 96–101, 106–7; overcompensation for 108–9; in pay 101–4, 113; of working time 92–6, 100–7, 108–10
flexible specialisation 70–2
Fordism 64, 70, 166, 240
Foster, Professor K. 26
France: research expenditure in 131; science parks in 5; scientists and technologists in 117, 122–3, 136, 137, 152, 246
freelance employees 101–2, 106
Freeman, C. 61
Fusfeld, H. 73

GEC Hirst Research Laboratories 119
gender 110–12
geography; of uneven development 164–5, 197–8; *see also* North–South divide; spatial considerations
Germany 61–2, 63, 84, 245; research expenditure in 131, 132; scientists and technologists in 117, 136, 137, 146, 245
Gorz, A. 116, 247
government 7–8; investment in science parks by 232, 236–8, 247, 249; and North–South divide 157–8; and research 65, 66–70, 119, 131–2, 157; (national laboratories 65, 67, 119, 131, 157, 171, 196, 233, 235); *see also* public sector
Grand Metropolitan 19, 23, 214
Greater London Council 246
Greece 137

Hall, P. 207, 241–2
Henneberry, J.M. 210, 225
Heriot-Watt Research Park 17, 19, 36, 37, 199–203
high-status employment: decline in 130–7; location of 156–8
history: industrial inheritance 177
Hoechst 61
Holford Report (1954) 171
house prices 155–6, 193

IBM 136
ICI 65, 119
image: price of 182; and property 215, 222–3; of scientists 147, 244–5; of small firms 142–3, 207; of work at science parks 86–91, 107, 113–14, 160–1, 206–9, 241, 250; *see also* symbolism
industrial estates 13–14
industry: counterposition of science parks with 87, 107, 113–14, 150–1, 160; regeneration of 197–203, 224–6, 237, 241–4, 248–9; and science 60–72, 161–2; *see also* academe–industry links; linear model; manufacturing industry
information technology (IT) 43–5, 140, 141; skills shortages in 151, 152, 153
inner city 23, 27, 163, 168
Inner City Partnership Scheme 19
innovation 56–85; alternatives to linear model of 79–85; diffusion of 58; process 77; product 77
Institute of Animal Physiology 171
international links 178–84; *see also* multinational corporations
investment 17, 19, 164, 178; in infrastructure 17, 209–26; return on 24–5, 29; rounds of 164, 209; *see also* local economy; private sector; public sector
Italy 71

Japan 71, 80, 84, 248; and British innovation 7; science parks in 5; scientists and technologists in 136, 146
Jeffers, Jeffe 23
job creation: as aim of science parks 28, 40–3, 168, 188–9

Kaplinsky, R. 80
Kelly, P. 78–9
King, Richard 207
Kline, S. 79, 81, 83
knowledge 73, 75–6, 80–1

labour: markets 187–94; (polarisation of) 188–9; shortages 146–7, 151–6, 188, 193–4, 246; *see also* division of labour; employment
larger premises, expansion into 41
Larsen, J.K. 206
Laser-Scan 35, 235
'leading-edge' technology 28, 43–50, 73–4
learning: by doing 73, 80; by using 80–2
Leeds Science Park 154
Levi, P. 206
linear model of innovation 9, 10–11, 52, 56–60; alternatives to 78, 84–5, 245–6; background to 60–72, 239–40; critique of 76–85; and employment 86–7, 119, 121, 128, 130, 135; and science parks 72–6, 158–9, 171–2
linkage: international 178–84; research 195–6; between science parks and local economy184–7
Little, A.D. 63
Lloyd, J. 90
Lloyds Bank 25–6, 168, 170
local authorities 217–21, 225, 236–8, 244; and economic strategies 235–8; industrial strategies investment 199–200, 243, 246–50; *see also individual councils*
local economy 11, 163–204; job creation within 28, 40–3, 168,

188–9; regeneration of 197–203, 224–6, 237, 241–4, 248–9
Lovell, Sir Bernard 68
Lucas 65, 166, 186

Macdonald, S. 78, 79
MacPherson, D. 89
Mallet, Serge 116
management: and career structure 117–18, 145; centralism of 64; education of 69; qualifications of 91–2, 145; status of 121; strategies of 24–8
Manchester Industrial Liaison Centre 32
Manchester Science Park 27, 32, 154
manufacturing industry: counterposition of science parks with 87, 107, 113–14, 150–1, 160; decline of 6–7, 8, 167; employment in 138, 140; in West Midlands 166–8
Marsh, P. 88
Marshall, M. 165, 198–9
Meacher, Michael 115
Medical Research Council (MRC) 45, 233
Merseyside Innovation Centre (MIC) 19, 23–4, 32, 244
Metropolitan-Vickers 65
Ministry of Agriculture 233
Ministry of Defence 233, 235
Ministry of Technology 68–9, 173
MIT (Massachussetts Institute of Technology) 72
monetarism 8, 150
Moreton, A. 191
Morrison, Herbert 67
Morse, Sir Jeremy 26
Mott, Sir Neville 68, 171–3, 174, 240
Mowery, D. 63–4, 65–6
multinational corporations 167, 179, 182–3; decision-making in 98, 183; and image of science park 182; self-definition in 160
multiplier effects 185
mystique *see* image

National Engineering Laboratory 119
National Institute of Agricultural Botany 171
National Physics Laboratory 119
national research laboratories 65, 67, 119, 131, 157, 171, 196, 233, 235
new start-up firms: academic 35–8, 72; as aim of science parks 27, 28, 30–4
New Tech 32
Newtech (Clwyd) Science Park 32
niche marketing 70–2
Nicholls, Harry 27
Noble, Harry 208–9
North–South divide 4, 12, 243; and academe–industry links 74; and aims of science parks 29; and employment 17, 19–20, 153–8, 175–6; in house prices 155–6, 193; location of science parks 16–17; and new start-ups 33–4; and patents 47; and public/private investment 19, 157–8, 208, 209–20, 226, 249–50; and rents 213–20; and size of firm 17
Nottingham Science Park 32
number of tenant firms 17

objectives *see* aims
organisational structure of firms 96–101, 106–7, 206–7
origins of science parks 5–6, 70
overtime pay 102, 103

Page, Ian 21–2
Parry, Dr Malcolm 23
part-time employees 112, 192
patenting activity 47–8, 49
Pavitt, K. 80
pay: flexibility of 101–4, 113; negotiation over 103–4; public-sector 131–2; of scientists and technologists 131–2, 134–5, 137; and skills 127
Pearson, R. 152–3
Peters, L. 73

Plant Breeding Institute 171
Plessey 14
Pollert, A. 106, 109
popular conceptualisation 9, 13–30, 51, 71–2; evaluation of 30–50
Portugal 137
post-Fordism 70–2
Poulantzas, N. 116
Prais, S.J. 146
Price, Christopher 23
private-sector investment 11–12, 19, 209–26 *passim*, 228, 229, 232, 235–8
product launches 48–9
production 45–6, 226–35; and R&D 204; science parks 226–35; separation from academe 85, 150–1, 157–62
profit-sharing schemes 102–3
property 177
property developments, science parks as 12, 213–26, 237
public–private sector relations 209–38
public sector 6; employment 119, 131–2, 157; *versus* entrepreneurship 206–9; investment in science parks 11–12, 17, 19, 170, 206–38, 247; markets 232; pay 131–2; status of 150–1

qualifications of employees 42, 46–7, 91, 111, 188; and career structure 122–3; of management 91–2, 145; and status 118–20, 121

recruitment policy 101, 107
Rees-Mogg, Lord 115
relational characteristics 244; of class and status 124–30 *passim*, 146; linear model of innovation 60; of spatial organisation 159; *see also* counterposition; exclusivity
relocation of firms to science parks 30–4, 42, 73
rents 182, 213–17, 222–3, 225–6
research 7, 8, 43–50; and academe–industry links 34–40; corporate 63–4, 65–6; employment in 131, 135–7, 152; (pay for 131–2, 134–5, 137); expenditure on 47, 48, 131, 136–7; and industry 60–72, 161–2; (*see also* linear model); and international links 181; public sector 65, 66–70, 131–2, 157–8, 233, 235; social aspects of 2, 59, 60, 120–1; spatial aspects of 58–9, 161, 204, 240–1; supply of ideas in 58; *see also* academe–industry links
research and development (R&D) 56–60, 63–5, 69–72, 119–21, 131–35, 233–5, 240–2
Research Associations 65, 67, 119
Rimmer, A. 23–4
risk, aversion of private sector to 221–2
Rogers, E.M. 206
Rogers, R. 170
Rose, H. and S. 62, 148
Route 128 (Boston) 5–6, 72
Rowe, D. 214, 215, 220–3
Royal Society 132, 133

sales and marketing activities 45–6
Scandinavia 80, 85
scientists and technologists: class and status of 115–62; (history of 116–21; recent changes in 130–62); *see also* academe–industry links; employment; research
Scottish Development Agency (SDA) 19, 28, 200–1, 217, 219
Segal, N. 176–7, 191, 226
self-employment 101, 141–2
Sematech 208
Senker, P. 136
Shaylor, G. 165, 168
Sheffield Science Park 244
Silberston, A. 134, 151
Silicon Glen 200
Silicon Valley 5–6, 26, 70, 72, 206, 208

skills and credentials 126–30, 140–1, 145–6; shortages of 146–7, 151–6, 188, 193–4, 246; *see also* qualifications
small firms 6, 8, 142–3, 148, 207; and local economy 182, 184; and North–South divide 17; research by 47
Snow, C.P. 3, 173
social considerations 2–3, 5, 9, 10, 11, 244–7; and academe–industry links 36, 40, 121, 196; of employment 3–4, 76, 115–62, 173–4, 189–93, 204, 246–7; and local economy 168–9, 177; and research 2, 59, 60, 120–1; and spatial aspects 2–3, 10, 160–2, 189, 240–1; *see also* class
South Bank Technopark 23
Spain 5, 137
spatial considerations 2–3, 9–10, 11, 158–62, 239–41; and academe–industry links 74–5, 159, 160, 241; and employment 3–4, 154–8, 189–94, 207–8; international networks 178–84; and public/private investment 209–26, 229, 232, 235–8, 249–50; and research 58–9, 161, 204, 240–1; and skills shortages 153–6, 193–4; and social aspects 2–3, 10, 160–2, 189, 240–1; symbolism 4, 10, 87; *see also* North–South divide
spatial separation of industrial activity from science parks 85, 157–62, 171–4, 204
specialisation, flexible 70–2
Stalker, G.M. 62–3, 120
Stanford Research Park 5–6, 26, 70, 72
start-ups *see* new start-up firms
state *see* government; public sector
status: job 76, 97, 111–13, 150–1; *see also* class; image
subcontracting 186
subsidiaries 31, 33; employment by 94, 95, 96, 98; and local economy 179, 182–3, 184
'sunrise', concept of 8, 150, 247–50
'sunset', concept of 8, 150, 247–50
suppliers, location of 179, 181, 185, 187
Surrey Research Park 17, 19; aims of 23; property of 214, 215
Swansea Science Park 37
Sweden 85
symbolism, spatial 4, 10, 87

tacit knowledge 80–1
'tapping-in' 38–40, 73
teachers, science and maths 146–7
technological change 82–3, 247–8
technological innovation *see* innovation
technology, level of: and aims of science parks 28, 43–50, 73–4; and international network 181; *see also* innovation
technology push 58
technology transfer 34–40; *see also* academe–industry links; innovation; linear model
Technology Transfer Institutes 202
Thomson, David 207
time of opening, unevenness of 14
trade unions 8, 104–5, 107, 113; white-collar 147–8
training 140–1, 146, 246
Trinity College (Cambridge) 24–5, 177–8, 190, 214, 217

unemployment: in Cambridge 175, 191; of graduates 153; in West Midlands 167–8, 188
Unilever 65
United Kingdom Atomic Energy Authority (UKAEA) 119, 131
United Kingdom Science Park Association (UKSPA) 13–14, 29–30
United States: and British innovation 7; research in 63–5, 66; (academe and 72–3, 75–6); science parks in 5–6, 30, 70, 72–3, 207, 208; scientists and technologists in 117

universities: employment in 130–3; investment in science parks by 217–19; *see also* academe–industry links
Urban Development Corporations 236

venture capital 36, 37, 232
Vincenti, W.G. 81
von Hippel, E. 81–2

warehousing 45–6
Warwick Science Park 19, 197, 199, 220–3
Watts, C. 24–5
Wavertree Technology Park 14
Welsh Development Agency 217, 219
West Midlands: economy of 165–8, 187–8, 196–9; *see also* Aston Science Park
West Midlands County Council 19, 220–1
West Midlands Enterprise Board 196–7, 198
West Midlands Technology Transfer Centre 187
West of Scotland Science Park 27–8
Whalley, P. 119–20, 121, 122–3, 142
Whyte, William H. 206
Wilson, Harold 8, 68, 148, 173
women, employment of 43, 110–11, 112, 149–50, 188, 191
Women and Work Training and Resource Centre 32
work *see* employment; labour
working hours 90–1, 102; flexibility of 92–6, 106–7, 108–10; recording of 95–6
Wright, E.O. 116–17, 122, 123–9, 144–5

Young (of Graffham), Lord 22–3, 133, 206
Yugoslavia 5

Zuckerman, Solly 173